American Institute of Chemical Engineers

Conference on Emerging Energy Frontiers in Research and Innovation

Topical Conference at the
2008 AIChE Spring National Meeting

April 6-10, 2008
New Orleans, Louisiana, USA

Printed from e-media with permission by:

Curran Associates, Inc.
57 Morehouse Lane
Red Hook, NY 12571
www.proceedings.com

ISBN: 978-1-60560-216-5

Some format issues inherent in the e-media version may also appear in this print version.

ISBN: 978-1-60560-216-5

American Institute of Chemical Engineers
Proceedings
Three Park Avenue
New York, NY 10016-5991
Phone: 212-591-8100

www.aiche.org

American Institute of Chemical Engineers

Conference on Emerging Energy Frontiers in Research and Innovation
2008

TABLE OF CONTENTS

26a Current and Upcoming Federal Policies Related to Biofuels and What Will Be the Impact

Jetta Wong, Environmental and Energy Study Institute (EESI), 122 C Street, NW, Suite 630, Washington, DC 20001

This paper will address major congressional policies related renewable liquid fuels and the impact these policies bring to Federal Programs. Major policies influencing renewable liquid fuels, i.e. the Farm Bill and the Energy Bill will be discussed and EESI perspective on future legislations that need to be implemented.

26b Next Generation Hydrocarbon Biorefineries

John Regalbuto, Chemical and Transport Systems Division National Science Foundation, National Science Foundation, Room 525.15, 4201 Wilson Blvd., Arlington, VA 22230

The recently passed Energy Independence and Security Act of 2007 mandates vastly increased production of renewable fuels at the same time as requiring a significant increase in fleet gas mileage. A good choice for renewable fuels is high energy-density hydrocarbons: green gasoline, green diesel, and green jet fuel. Recent cutting edge research has demonstrated a number of feasible pathways for the conversin of lignocellulosic biomass into these renewable hydrocarbon fuels. These pathways will be reviewed, and the interagency efforts to direct new research on biomass conversion will be described.

26c Challenges to a Successful Biofuels Industry and Department of Energy Efforts to Overcome Them

Valerie Sarisky-Reed, Office of the Biomass Program, US Department of Energy, 1000 Independence Ave, SW, EE-2E, Washington, DC, DC 20585

This paper will address the technical, political, and infrastructure barriers facing the development of a successful biofuels industry in the U.S., as well as the efforts the Department is taking to overcome them through its Biofuels Initiative. This discussion will include a description of the goals of the Biofuels Initiative, current programmatic work, major research and development successes, and planning for the future.

26d Biomass to Biofuels, Strategies and Advances in Basic Science to Overcome Major Research Barriers

Raul Miranda, Office of Basic Energy Science, US Department of Energy, Office of Science, SC-22.1/ Germantown Building, 1000 Independence Ave, SW, Washington, DC 20585-1290

This paper will describe the report titled "Basic Research Needs: Catalysis for Energy" from the U.S. Department of Energy, Office of Basic Energy Sciences Workshop August 6-8, 2007, in Bethesda, Maryland.

POWER TRANSFER AND INDUCTANCE IN A STAR CONNECTED 3-PHASE RC CIRCUIT AMPLIFIER

Himanshu Dehra, American Institute-Industry Forum on Energy, Lachine, Québec, Canada

Introduction

Amplifiers convert input signal of one type of voltage, current or power into an output signal of another type of voltage, current or power using a certain amplification factor. The most common application of an amplifier is in converting a low-voltage signal from a compact disc player to a level reachable for operation of a pair of loud speakers. The overall operational characteristics and frequency response of the system are determined in the system with source-a compact disc player, circuit-an amplifier and connected load-the loud speakers. With use of filters, it is also possible to reduce or eliminate signals of unwanted frequencies, such as those caused by electrical noise or other types of noise interference. ***The aim of the paper is to analyse a parallel plate photovoltaic device connected to a potentiometer as a star connected 3-Phase Resistance-Capacitance (RC) circuit amplifier***. The effect of inductance and resulting power transfer has been determined in the RC circuit amplifier constituting of a parallel plate photovoltaic device. The analysis has been performed on the basis of the accepted unified theory for stresses and oscillations, as proposed by the author (Dehra, 2007a). The brief on the electrodynamics of a pair of photovoltaic (PV) modules with connected resistance due to thermal mass of an outdoor room has been previously presented (Dehra, 2007b). ***The analysis has also discussed from the electrodynamics point of view, power transfer and effect of induction losses in a 3-Phase RC circuit amplifier constituting of a parallel plate photovoltaic device***. The theory of the sinusoidal steady-state response was applied in performing the analysis of the circuit, because of the advantage of representing a periodic function in terms of a sinusoidal exponential function.

A Star Connected 3-Phase Power Generation

The full scale experimental setup for a parallel plate photovoltaic device connected to a potentiometer was installed in an outdoor room facility located at Concordia University, Montréal, Canada (Dehra, 2007c). The schematic of the experimental setup is illustrated in Fig. 1 (Dehra, 2007c). An amplifier was built with a pair of glass coated photovoltaic (PV) modules forming a parallel plate duct with a plywood board and connected to a potentiometer. A potentiometer, a wire-wound variable resistor of up to 50 Ω was a wire-wound circular coil with a sliding knob contact (Dehra, 2007c). It was used to vary electrical resistance across connected PV modules without interrupting the current. A star connected RC circuit amplifier constituting a parallel plate photovoltaic device connected to a potentiometer was constructed. A star connected 3-phase parallel plate photovoltaic device was a pair of glass coated PV modules (1st phase, 1st line), air cavity (2nd phase, 2nd line) and polystyrene filled plywood board (3rd phase, 3rd line). The neutral (4th line) of the 3-phase connection was a wooden base of the parallel plate photovoltaic device installed in a pre-fabricated outdoor room. The construction of the star connected 3-phase power generation and supply has been depicted in Fig. 2. The circuit diagram for the circuit has been provided in Fig. 3. The expressions for instantaneous power transferred to an impedance load for a three phase power generation from PV modules, air and polystyrene filled plywood board are presented as:

$$P_a(t) = V_p I_p \cos(\theta) + V_p I_p \cos(2\omega t - \theta) \tag{1}$$

$$P_b(t) = V_p I_p \cos(\theta) + V_p I_p \cos(2\,\omega\, t - 2\,\alpha - \theta) \tag{2}$$

$$P_c(t) = V_p I_p \cos(\theta) + V_p I_p \cos(2\,\omega\, t - 2\beta - \theta) \tag{3}$$

Where V_p and I_p are the rms values of voltage and current, respectively, θ is the phase angle between them, α is the angle between Phase a (PV module) and Phase b (Air cavity), and β is the angle between Phase c (plywood board) and Phase a (PV module).

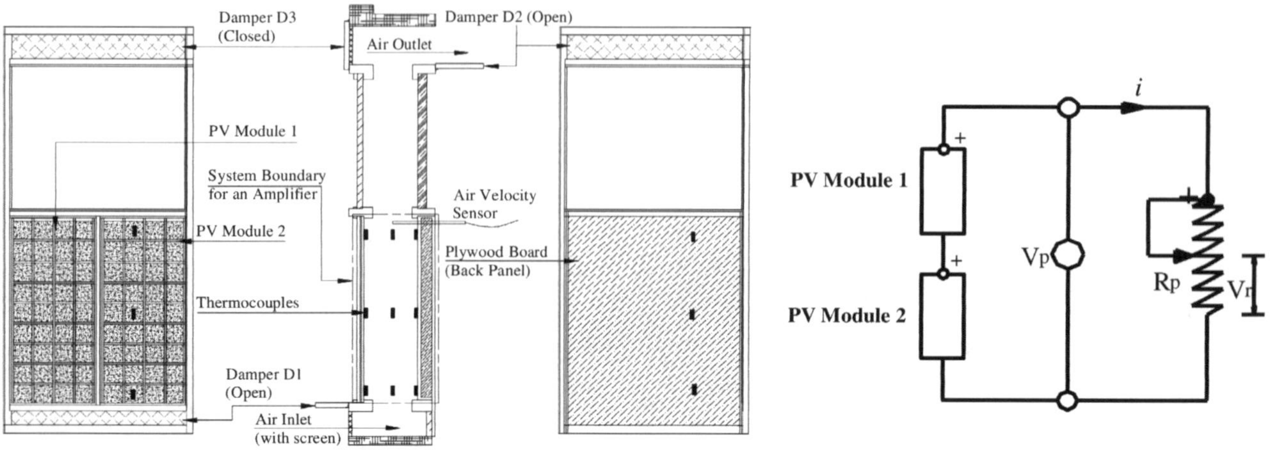

(a) *(b)*

Figure 1: Schematic of experimental setup for a parallel plate photovoltaic device connected to a potentiometer: (a) location of sensors; (b) Electrical circuit diagram

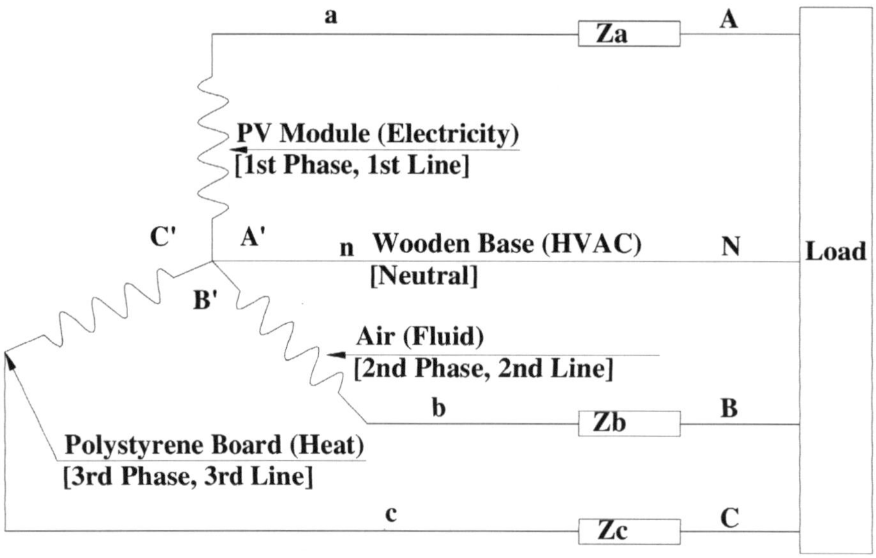

Figure 2: A Star Connected 3-Phase Generation using a Parallel Plate Photovoltaic Device

A Parallel Plate Photovoltaic Device

The characteristics of a parallel plate photovoltaic device connected to a potentiometer were established by varying electrical resistance with rotation of knob of a potentiometer. The current-voltage measurements were obtained for determining electric power output with a series electrical circuit connection of a pair of vertically inclined PV modules installed on a wooden frame. The electrical measurement results of currents, voltages and power with varying electrical resistance of potentiometer are presented in Table 1. The results of the power output from a potentiometer with rotation of circular knob are illustrated in Figure 3. The phenomenon of photovoltaic amplification has been observed from the graph of Figure 3. The gain in steady state electrical for a photovoltaic device is a factor of its volume or resistance. This operational characteristic is similar to operation of a loudspeaker.

Table 1. Sample electrical measurement results with varying resistance of potentiometer

Rotation	Volts	Amps	Watts	Rotation	Volts	Amps	Watts	Rotation	Volts	Amps	Watts
240°	18.7	-	-	83°	16.3	0.935	15.23	30°	9.7	1.577	15.21
239°	16.5	0.331	5.461	75°	16.0	1.014	16.26	27°	9.0	1.587	14.33
201°	17.4	0.414	7.195	69°	15.8	1.100	17.38	21°	7.1	1.583	11.24
185°	17.5	0.454	7.940	64°	15.5	1.165	18.04	18°	6.2	1.573	9.831
162°	17.3	0.513	8.885	55°	15.0	1.302	19.53	17°	5.7	1.578	9.026
150°	17.18	0.550	9.449	50°	14.5	1.386	20.05	12°	3.9	1.567	6.257
142°	17.19	0.582	10.00	43°	13.2	1.503	19.79	10°	3.2	1.553	4.840
128°	17.1	0.640	10.93	42°	13.1	1.493	19.49	1.5°	0.5	1.593	0.807
107°	16.8	0.750	12.51	37°	11.9	1.536	18.26	1°	0.3	1.59	0.426
89°	16.4	0.884	14.45	32°	10.5	1.567	16.42	0°	-	1.643	-

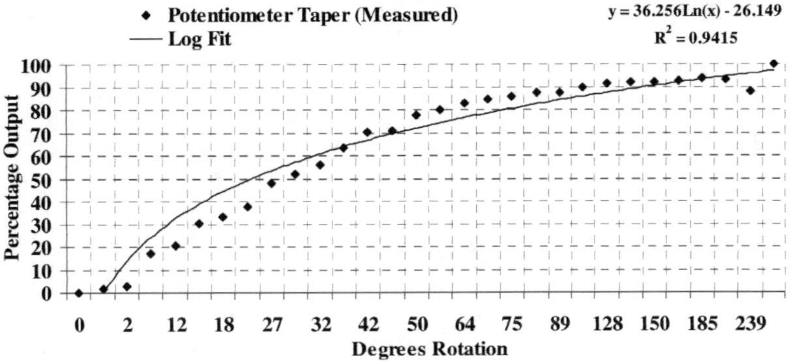

Figure 3: Potentiometer taper as function of percentage voltage output

Electrical Parameters for a RC Circuit Amplifier

Capacitance: The capacitance of a parallel plate photovoltaic device with air as a dielectric medium was calculated to be 91.2 picofarads.

Resistance: The electrical resistance of various components were calculated as: glass coated PV modules were approximated as 5.3 kΩ, air was approximated as 1200 MΩ, and plywood board was approximated as 26.5 Tera Ω. The total equivalent electrical resistance of a parallel plate photovoltaic device was approximated as: 5.3 kΩ.

Time Constant: The time constant, which is product of resistance and capacitance, was calculated to be: 0.5 microseconds. The frequency with this time constant was calculated to be 2 MHz.

Capacitive Reactance: The capacitive reactance was calculated to be: 872.5 Ω.

Impedance: The impedance of the circuit was calculated to be: 5.4 kΩ.

The Phase angle θ: The phase angle between capacitance and reactance was calculated to be 9°.

***The Phasor representation*:** $Z = 5.300 - j\,0.8725 = 5.4\ k\Omega\ \angle -9°$.

***Capacitive Heating*:** The joule law gives instantaneous power absorbed by the capacitive impedance and is converted to heat. The heat capacities under critical operation of buoyancy-induced hybrid ventilation were calculated to be 59.6 kJ, 0.755 kJ and 510.7 kJ for PV module, air and plywood board respectively (Dehra, 2007 c). The total average value of joule heating for the parallel plate photovoltaic device was calculated to be 571 kJ.

Induction Losses: The induction losses due to thermal storage effect in the parallel plate photovoltaic device was calculated to be 15.9 KJ (Dehra, 2007c).

Power Factor: The power factor was calculated to be $\cos\theta = 0.911$ lag.

***Current function* ($i^2(t)$):** Using the current function, $i^2(t) = I_m^2 \sin^2(\omega t + \theta)$, the effective (root mean square) value of current was calculated to be 10.4 amps and maximum value of current was calculated to be 14.71 amps.

Voltage function: The voltage function is defined as per the sine wave: $v = V_m \sin(\omega t)$. The effective value of the voltage was calculated to be 60.4 volts and maximum value of the voltage was calculated to be 85.42 volts.

Power function: The instantaneous power is given by the expression:

$$p(t) = \frac{V_m I_m}{2}\cos(\theta) - \frac{V_m I_m}{2}\cos(2\,\omega\, t - \theta) \tag{4}$$

The Plots: The time diagram for current, voltage is plotted in Figure 4(a). The time diagram for power is plotted in Figure 4(b).

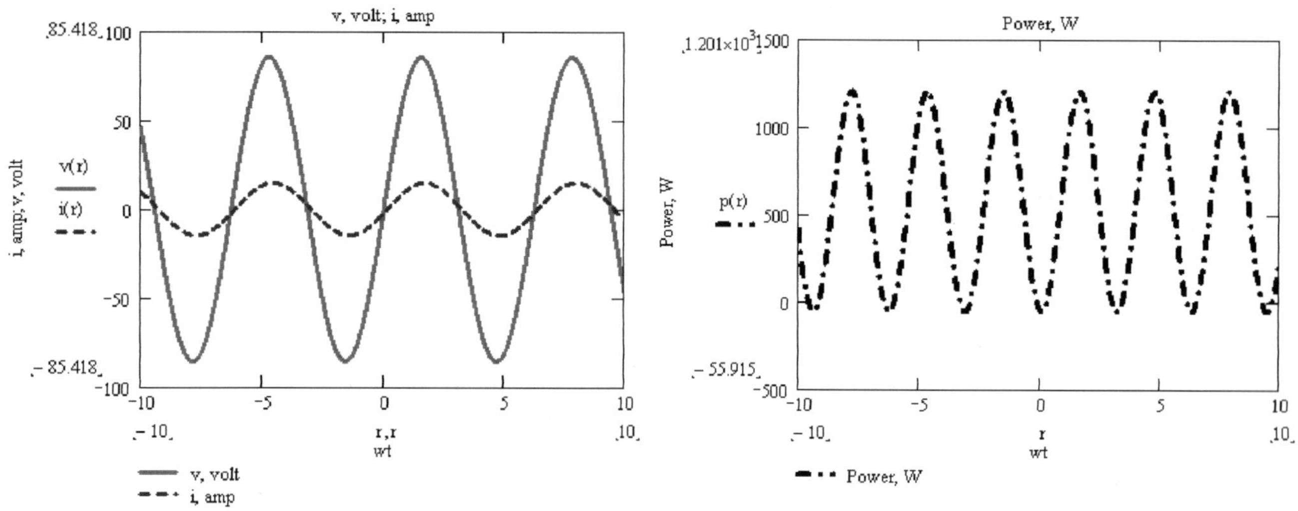

Figure 4: Time diagrams: (a) voltage and current; (b) power in a RC circuit amplifier

Power Transfer and Effects of Inductance

Capacitive Reactance and Resistance in Series: The losses that appear in capacitive circuits are lumped in a resistor connected in series with the capacitor.

Capacitance and Resistance in parallel: When a sine waveform voltage is applied across a capacitor and a charging current of sine waveform is transmitted across the circuit. The alternating current source, which is a sine waveform voltage, if applied at a uniform rate, is responsible sinusoidal response of the charging current in the capacitor. The motion of the charging current is transmitted through the capacitor, corresponds to the electron flow in the wires connecting the capacitor to the alternating current source. The alternating current source is responsible for development of the electric stress in the dielectric between the plates of the capacitor. Electron flow does not occur through the capacitor. Instead, the electrons flow around the capacitor circuit on one alternation causing a negative charge to build up on one place, and a corresponding positive charge on the other, and the next alternation causing a reversal of the polarity of the charges on the plates. Thus, the effective impedance which the capacitor offers to the flow of alternating current can be relatively low at the same time that the insulation resistance which the dielectric offers to the flow of direct current is extremely high.

Power Transfer: With no voltage or charge, the electrons in the dielectric between the capacitor plates rotate around their respective nuclei in normally circular orbits. When the capacitor receives a charge the positive plate (PV Module) repels the positive nuclei and at the same time the electrons in the dielectric are strained toward the positive plate and repelled away from the negative plate (Plywood Board). This distorts the orbits of the electrons in the direction of the positive charge. During the time the electrons are changing from normal to the strained position there is a movement of electrons in the direction of the positive charge. The movement constitutes the displacement current in the dielectric. When the polarity of the plates reverses, the electron strain is reversed. If a sine-wave voltage is applied across the capacitor plates the electrons will oscillate back and forth in a direction parallel to the electrostatic lines of force. Displacement current, is a result of the movement of bound electrons, whereas conduction current represents the movement of free electrons.

The Figure 4(b) shows that the instantaneous power is negative whenever the voltage and current are of opposite sign. However, as has been illustrated in the Figure 4(b) that positive area of p(t) energy exceeds the negative area. Therefore, the average power is finite. Since the angle, θ, is small between cuurent and voltage, the negative area of p(t) energy become very small. During the first quarter cycle (from 0° to 90°) the applied voltage rises from slightly negative value to a maximum and the capacitor is receiving a charge. The power curve is positive during this period and represents energy stored in the capacitor. From 90° to 180°, the applied voltage is falling from maximum to slightly negative value and the capacitor is discharging. The corresponding power curve is negative and represents energy returned to the circuit during this interval. The third quarter cycle represents a period of charging the capacitor and the fourth quarter represents a discharge period.

Induction Losses: The induction losses due to thermal storage amount to 1.5% in comparison to the capacitive heating. When a circuit containing a coil or source of energy is energized with direct current, the coil's effect in the circuit is evident only when the circuit is energized, or when it is de-energized. However, when the inductive circuit is supplied with alternating current, the induction losses are continuous and much greater than when it was supplied with direct current. For equal applied voltages, the current through the circuit is less when alternating current, is applied than when direct current is applied. The alternating current is accompanied by an alternating magnetic field around the area of the source of energy, which cuts through the area of the source of energy in the circuit. Most of the applied voltage appears across inductance, L, with little remaining for the load in the circuit. In a circuit possessing inductance only, the true power is zero. The current lags the applied voltage by 90°. The areas of induction losses above the X axis represent positive energy and the areas below the X axis represent negative energy. Thus induction losses can not be avoided in any electrical circuit, but can be minimised.

Applications

The amplitude of an electricity wave is defined as the power storage per unit area per unit time.

The electrical power is stored in a packet of electricity wave of unit cross sectional area and of length s, the speed of light. Therefore, the electrical power intensity I is the product of total power storage capacity for a packet of electricity wave and the speed of light. The logarithm of two electrical power intensities, I_1 and I_2, gives power difference for two electrical power intensities. It is mathematically expressed as:

$$Sol = \log \left(I_1\right) \left(I_2\right)^{-1} \tag{5}$$

Where, Sol is a dimensionless logarithmic electrical noise unit. It is also a unit for solar noise. The decisol (dS) is more convenient for electrical power systems. Since a decisol (dS) is $1/11^{th}$ unit of a Sol, it is mathematically expressed by the equation:

$$dS = 11 \log \left(I_1\right) \left(I_2\right)^{-1} \tag{6}$$

The base of logarithm is to be determined by the designed rating of the electrical power system. This has to be done in conjunction with design parameters for the reference rating system for defining an electrical noise. The above definition will be extremely useful for estimating accurate performance rating of electro-optical devices such as photovoltaic modules, light-emitting-diodes etc. It will also definitely bring new design parameters for determining the load on an amplifier circuit, as per the designed rating of electro-magnetic energy transducers. The design parameters are to be very critical at occurrence of the acoustic resonance (Dehra, 2007a).

Conclusion

The analysis has been performed for power transfer and inductance in a star connected 3-phase resistance and capacitance (RC) circuit amplifier. An unbalanced 3-phase amplifier constituting a parallel plate photovoltaic (PV) device connected to a potentiometer was constructed for the purpose of analysis. A parallel plate photovoltaic device was installed in a prefabricated outdoor room. The electrical and thermal properties and measurements were collected to analyse unbalanced power transfer and inductance in a non linear RC circuit amplifier. The analysis has been conducted on the basis of accepted unified theory for stresses and oscillations. The industrial applications of an innovative photovoltaic amplifier are in defining and designing of electrical noise, electro-optics and electro-magnetic energy transducers.

References

1. Dehra, H., (2007a), "A Unified Theory for Stresses and Oscillations", Canadian Acoustics, September 2007, Vol. 35, pp.132-133.
2. Dehra, H., (2007b), "The Electrodynamics of a Pair of PV Modules with Connected Building Resistance", in the proceedings of The 3rd IASME /WSEAS International Conference on ENERGY, ENVIRONMENT, ECOSYSTEMS and SUSTAINABLE DEVELOPMENT (EEESD'07), Agios Nikolaos, Crete Island, Greece, July 24-26, 2007, pp. 562-565.
3. Dehra, H., (2007c), "The Effect of Heat and Thermal Storage Capacities of Photovoltaic Duct Wall on Co-Generation of Electric and Thermal Power", in the proceedings of American Institute of Chemical Engineers - AIChE 2007 Spring Meeting, Houston, Texas, USA, April 22-26, 2007, session 36a.

Appendix: Definitions

Acoustic Resonance: Acoustic resonance is defined at maximum absorption of energy, when frequency of oscillations matches with the natural frequency of oscillations from a particle body system. The oscillations are assumed to be propagated at their maximum absorbed energy due to transmission of light, sound, noise, heat, electricity, fluid and fire from a system propagating at its natural frequency of oscillation.

Capacitance: Capacitance is defined as the property of an electrical device or a circuit that tends to oppose a change in voltage. Capacitance is also a measure of the ability of two conducting surfaces, separated by some form of non-conductor, to store an electric charge.

Climate-Particle Oscillations: The oscillations of fluid particles present in Earth's climate are under the influence of electro-magnetic and gravitational forces of a solar system. Examples of climate-particle oscillations are wind and tides. The cyclic tides occur due to electro-magnetic and gravitational forces of Sun and Moon acting on the Earth. The random oscillations of wind occur due to rotation of the Earth and differential heating of the Earth's atmosphere between its equator and poles.

Cyclic Load: The stresses on a particle in a system are caused due to presence of cyclic electro-magnetic load of day/night, heating/cooling and compression/expansion applied on a particle body. The steady state normalized cyclic load on vertical members of the system is obtained from the inclined incident load.

Electrical Analog: The use of electrical A.C. circuit theory is often made to solve problems of Acoustics. The vertical and horizontal members of a particle body system are composed of distributed electrical elements developed due to electro-magnetic forces. Table 2 has presented electrical analogs for mechanical, hydraulic and thermal systems.

Inductance: Inductance is the characteristic of an electrical circuit that makes itself evident by opposing the starting, stopping, or changing of current flow. Inductance is the characteristic of an electrical conductor which opposes a change in current flow. Even a perfectly straight conductor has some self inductance.

Stresses on a Particle Body System: The stresses developed on a particle due to various forces are classified as: i) fundamental stresses; ii) internal stresses; and iii) external stresses. The fundamental stresses are developed due to presence of gravitational and electro-magnetic forces of a solar system. The internal stresses are developed under the influence of fundamental stresses and are defined by properties of composition of a particle. The external stresses are developed under an external source of energy. The oscillations of a particle are assumed to be generated because of stress development with cyclic load of day/night, heating/cooling and compression/expansion.

Table 2 Electrical Analogs

	Mechanical System	**Electrical Analog**	**Hydraulic System**	**Electrical Analog**	**Thermal System**	**Electrical Analog**
1.	Force	Current	Flow rate	Current	Heat	Charge
2.	Velocity	Voltage	Fluid level	Voltage	Heat flow	Current
3.	Mass	Capacitance	Fluid Resistance	Electrical Resistance	Temperature difference	Voltage
4.	Compliance	Inductance	Fluid storage	Capacitance	Heat capacity	Capacitance
5.	Viscous damping	Conductance	Fluid volume	Charge	Thermal resistance	Electrical resistance

96c High Temperature Co-Electrolysis of Steam and Carbon Dioxide to Produce Synthesis Gas Utilizing Concentrator Photovoltaic Arrays

Lyman Frost[1], Joseph Hartvigsen[2], and S. Elangovan[2]. (1) Energy and Electrochemistry, Ceramatec, Inc, 2425 South 900 West, Salt Lake City, UT 84119, (2) Electrochemistry, Ceramatec, Inc, 2425 South 900 West, Salt Lake City, UT 84119

High Temperature Co-electrolysis of Steam and Carbon Dioxide to Produce Synthesis Gas Utilizing Concentrator Photovoltaic Arrays

Ceramatec, Inc is currently developing technology for the high temperature co-electrolysis of steam and carbon dioxide to produce synthesis gas. This technology utilizes solid oxide fuel cell materials to produce a clean stream of carbon monoxide and hydrogen. By varying the concentrations of carbon dioxide and steam, the yielded ratio of hydrogen to carbon monoxide can be modified dependent on the type of synthetic fuel desired for production.

High temperature electrolysis is substantially more efficient than existing electrolysis methods since part of the energy required to break the molecular bonds is furnished by heat. Existing proton exchange membrane electrolysis technologies operate below 100O C but high temperature electrolysis operates in the 800O C range. This higher temperature improves the efficiency significantly. The use of the materials utilized for solid oxide fuel cells matches the desired temperature range. Issues with the materials are raised, however, since the operating conditions of the electrolysis cells vary from those of solid oxide fuel cells. While some environmental conditions are beneficial, others create potential problems that Ceramatec is addressing with its research.

Ceramatec has been working on application of this high temperature technology for production of synthesis gas for production of synthetic methane for a company in Australia using solar concentrator technology as a source of heat and electricity. The Office of Naval Research has directly funded Ceramatec to investigate the production of synthetic fuels through the co-electrolysis of carbon dioxide and steam.

This paper will present results of laboratory scale high temperature co-electrolysis. The paper will present the yields of hydrogen and carbon monoxide obtained from varying input mixtures of carbon dioxide and steam. The paper will also present the results of utilizing the resultant synthesis gas to produce synthetic methane and Fischer Tropsch liquid fuels. Issues related to long term lifetime testing, commercial scale-up, and cost will be discussed.

97a Hydromax - Breakthrough Technology for Conversion of Biomass to Fuels and Chemicals

Jerry Stephenson, Chief Systems Engineer, Diversified Energy Corporation, PO Box 1239, Gilbert, AZ 85233

HydroMax is a patented, breakthrough gasification technology that utilizes a unique chemical pathway to produce syngas (CO + H2) and hydrogen (H2) in separate and distinct streams, resulting in process and economic benefits. Using two distinct steps, the HydroMax process begins with a molten Iron/Tin (FeSn) bath heated to 1300°C. In Step 1, steam is injected into the molten bath, which is then thermo-chemically split (i.e. traditional "steaming" of iron) resulting in H2 gas (released) and oxidized Fe (FeO). After the Fe is oxidized, steam injection ceases and biomass is injected into the reactor (Step 2). Carbon, with its high affinity to oxygen, reduces the oxidized FeO to its pure form (Fe) and produces a CO-rich syngas, which is either used as a fuel gas or combined with hydrogen from Step 1 to produce Fischer Tropsch fuels. Based on these fundamental processes of steaming-iron and metal-bath-smelting, the oxidation and reduction steps are then repeated in a cyclic fashion thus resulting in no molten-metal consumption. The process can also be operated in a continuous mode where both steps occur simultaneously if the feedstock moisture content is high. Furthermore, the tin (Sn) in the molten metal serves to capture sulfur as tin-sulfide gas that is subsequently 'roasted' to tin-dioxide and reintroduced into the molten-metal bath, therefore maintaining the tin composition integrity and operating temperature of the reactor.

Several unique attributes of the HydroMax gasification technology result in significant operational and economic benefits compared to existing gasification approaches, including:

• Capital cost reductions as a result of a simple and proven reactor design.

• Smaller gasification footprint due to the high thermal inertia of the molten-metal reactor.

• Flexibility to utilize any carbon feedstock due to the high thermal inertia associated with a molten-metal bath.

• Flexible and tailored outputs as a result of the option to operate in cyclic or continuous mode.

• Ability to accept high sulfur feedstock due to the unique sulfur removal approach via tin-sulfide.

Commercial and industrial HydroMax applications are numerous, including: utility-scale power generation, synthetic natural gas production, biomass-to-liquid-fuels, and large-scale hydrogen production for oil refineries or the fertilizer industry. Currently, the HydroMax team is pursuing an initial market entry point focused on small industrial scale applications for converting biomass into a fuel-gas for process heating and drying applications used by a number of manufacturing companies.

Diversified Energy is cooperating with two industrial firms, in different markets, to design a HydroMax gasification system to supply a natural gas substitute for process heating and to explore the economic feasibility of incorporating HydroMax as a source for low-cost fuel gas.

Previous laboratory and proof-of-concept tests have successfully demonstrated the fundamental science and approach of the HydroMax technology. Capital cost analysis conducted by Aker Kvaerner indicates that significant cost reductions are achievable using the HydroMax approach. Since this analysis, the HydroMax technology team has made significant development progress. A dynamic simulation model was created to assess the technology performance using a variety of biomass feedstock types. Initial results from this model indicate that HydroMax performs better than existing commercial gasifiers, particularly when using low-rank biomass sources as a feedstock. Also since 2006, HydroMax was selected by the Department of Energy for Phase I SBIR funding, which will be used to perform a suite of bench-scale tests using multiple feedstocks. The HydroMax team has also developed multiple operational concepts in concert with several potential clients including a pulp mill and gypsum drywall manufacturer.

The AIChE presentation will include: 1) A brief overview of the HydroMax gasification technology 2) An introduction to the HydroMax development team 3) Previous test results and analyses including recent test results from DOE funded tests scheduled for January 2008 4) Analyses conducted, and outputs generated, using models developed from AspenPlus® and a discussion of HydroMax performance compared to commercially available gasifiers 5) A brief summary of the industrial applications being pursued and evaluated as candidate HydroMax entry applications including notional plant layouts and interfaces 6) Summary results from detailed economic analyses performed and 7) Future development plans and activities.

Web Page: www.diversified-energy.com

97b H2 Purification and CO2 Capture in a Post-Biomass-Gasifier Membrane Reactor

Charles Taylor, Bryan D. Morreale, and Bret H. Howard. U.S. DOE/NETL, P.O. Box 10940, Pittsburgh, PA 15236-0940

NETL envisions that the gasification of carbonaceous feedstocks may be the near- to mid-term sources of hydrogen for the transition to a renewable, hydrogen-based economy. However, the environmental impacts associated with the generation and emission of greenhouses gases from the gasification process remains a substantial concern. Therefore, NETL has devoted substantial resources towards the identification of efficient hydrogen separation and carbon capture/sequestration technologies. Hydrogen membranes integrated into a water-gas shift membrane reactor have been identified as a promising means of maximizing the production of pure hydrogen while simultaneously yielding a high-pressure, concentrated CO2 containing stream ready for sequestration.

In the gasification process the carbonaceous feedstock (biomass, coal, petroleum resid, etc.)is converted to "synthesis gas" in the presence of steam and oxygen. The gasifier effluent, which is primarily composed of CO, CO2, H2, H2O and contaminants such as H2S and NH3, can exhibit temperatures and pressures as high as 1000 C and 7 MPa, respectively. The high-pressure, high-temperature gasifier effluent stream then proceeds through various processes, each process operating at a lower temperature than the previous. These process steps are used to remove particulates and sulfur compounds and further shift the CO-rich effluent stream to H2 and CO2 via the water-gas shift reaction. Finally the potential for hydrogen separation, carbon dioxide capture and power generation can be realized.

Process simulation studies have shown that the integration of membrane technologies into the gasification process can both increase efficiency and reduce the cost of hydrogen production while yielding a CO2 stream ready for sequestration. However, for this integration to be realized, viable membranes must be identified that exhibit the required permeability, selectivity, chemical resistance and mechanical stability.

Research groups, including NETL, are exploring the viability of dense metal hydrogen membrane technologies to enhance the production and separation of hydrogen as well as the capture of carbon dioxide from the coal gasification process. Areas of research to ensure the success of dense metal membrane technologies for implementation into the gasification scheme includes the fabrication of thin metallic films and support materials, an understanding of the interaction of membrane materials in the presence of both major and minor gas constituents, chemical and temperature induced morphological changes and the identification of new membrane materials through experimental and computational exploration.

Fast Pyrolysis of Poplar Wood and Poultry Litter

Sedat H. Beis and Foster A. Agblevor. Biological Systems Engineering, Virginia Polytechnic Institute and State University, Seitz Hall, Blacksburg, VA 24061

1. Introduction

Rising global economic, strategic and environmental concerns putting pressure on the use of more renewable energy options. Among the few renewable energy sources biomass is the only renewable carbon source that has potential applications as fuels and chemicals. Majority of the biofuels such as bioethanol and biodiesel compete with food for raw material. This situation will create major supply and cost concerns in the future. On the other hand, biomass may become better feedstock alternatives for the production of biofuels and chemicals with potential advantages in ready availability. Pyrolysis technology is one of the thermochemical conversion method for producing solid, liquid and gas products in the utilization of biomass. Fast pyrolysis is particularly important for bio-oil production that can be further processed into fuels and chemicals [1,2,3].

2. Materials and methods

After air drying to moisture level to 8-10% the poplar wood was ground in a cutting mill and poultry litter was ground in an hammer mill. Each feed was sieved through 16 mesh prior to pyrolysis. Bulk densities of the samples were determined by measuring the mass corresponding 500 mL of sample in a measuring cylinder without packing.

Pyrolysis of the each sample was carried out in a bench-scale fluidized bed pyrolysis system. The fluidized bed reactor used was described in a previous paper [4]. 100 g Silica sand (-28 +60 mesh) was used as the heating medium and the bed was fluidized with nitrogen. A metal plate having 100-µm porosity used as gas distributing media and nested the heating media. The reactor was externally heated with a three-zone electric furnace. Biomass samples were fed into a feed hopper and conveyed by a twin-screw feeder into an entrainment zone where high-velocity nitrogen gas entrained the sample into heart of the fluidized bed. The reactor tube contained a bubbling fluid bed with back-mixing of the feed and sand. The static sand bed height was 38 mm, expanding to 100 mm. when fully fluidized. Nitrogen flow rate was set to $2V_{min,f}$. The pyrolysis temperature was maintained at 450-475 °C. Runs lasted from 90 to 360 min and the feed rate was 100-360 g/h. The feed rate, gas flow rate, and reactor temperature were kept constant during each run. The nitrogen and pyrolysis gases and vapors exiting the reactor passed through an externally heated hot gas/vapor filter to separate char/ash and any entrained sand. The hot gas filter temperature and transfer lines were maintained at 400 ^{o}C to avoid condensation of the pyrolysis vapors before they reach the first condenser. The pyrolysis gases and vapors were then passed through a cooling system consisting of two double pipe condensers, connected in series and maintained at -5 ^{o}C by circulating chilled water. An electrostatic precipitator and an coalescent filter were used for trapping the effluent leaving the 2nd condenser. The electrostatic precipitator was maintained at 20 kV.The amount of the liquid and solid products were determined by measuring the weight gain in the corresponding unit after the system cooled down to room temperature. Gas product yields were calculated by difference.

The proximate and ultimate analysis of the poplar wood, poultry litter and their bio-oils and chars were determined by Galbraith Analytical Laboratory (Knoxville, TN, USA). Viscosity measurements of the oils were recorded by using Brookfield DV-II+ PRO Digital Viscometer. Spindle SC4-18, 8 mL sample chamber and the thermosel system were used in the measurements. pH measurements were determined with a Corning pH meter 440.

The molecular weight distribution of the poplar wood and poultry litter biooils were determined by GPC. Biooil samples (~80 mg) were dissolved in BHT inhibited THF (20.0 mL). After 24hr dissolution, the solutions were filtered through a 0.45 mm Teflon syringe filter. Eight polystyrene standards with peak molecular weights of 22000, 7000, 5050, 3570, 1700, 1050, 510, 162 (supplied by American Polymer Standards) were used to construct a universal calibration curve. Approximately 100 µL aliquots of each standard and the samples were individually injected through a 20 µL Rheodyne 7125 injection port for analysis. Separation was performed using Waters HR1 and HR3 columns connected in series. The THF mobile phase containing 0.025 wt% BHT inhibitor was pumped at 1.0 mL/min from a Waters 515 HPLC pump connected to a Li-ChromeIII flat coil pulse dampener. Inline solvent degassing was performed with a Phenomenex DG4000 vacuum degasser. A viscotek TDA 302 Triple array detector consisting of right angle and low angle light scattering detectors, an inline viscometer and a differential refractometer was used for signal detection. Viscotek Omnisec 3.1 Software was used for data analysis.

^{13}C NMR spectra of bio-oils were recorded in DMSO-d_6 solutions at 100.6 MHz and 25°C using a VARIAN 400 MHz spectrometer .Solutions of 50 % (w/w) bio-oil were employed in 5 mm diameter tubes. About 32768 scans were accumulated for each sample. The ^{13}C spectra were divided into carbonyl, total aromatic, carbohydrate and methoxy/hydroxy-bound carbon shift ranges for integration [5].

The gaseous products were analyzed by gas chromatography using an on-line SRI Multigas#2 GC instrument. High purity He was used as carrier gas 20mL/min in both of the auto sampling loops which were connected to Hayeseep-D 6' column, MXT-1 60 m capillary column and molecular sieve-13X 3' column. TCD and methanizer coupled FID detectors were used for detection.

Gas stream were analyzed online for determination of major gaseous products mainly CO_2, H_2, CO, CH_4, C_2H_2, C_2H_4, and C_2H_6.

3. Results and discussion

The average liquid yield for the poplar wood was 60±0.2% and that for the broiler litter was 40±1.1% on as received biomass basis. The char and gas yields for the poplar were 12±2.1 % and 28±2.3% and that for the broiler litter were 35±2.7% and 25±2.9% respectively. The ultimate and proximate analysis data of the poplar wood, poultry litter and their bioils are given in Table 1. The poplar biooil and broiler biooil contained 57.80% and 62.84% carbon, 6.80% and 8.31% hydrogen, <0.5% and 7.23% nitrogen, <0.05% and <0.9% sulfur respectively. The pH, viscosity and high heating value for poplar biooil and broiler biooil was 3.0 and 6.5; 320cP (45°C) and 423cP (60°C); 22.48 and 27.98 MJ/kg respectively.

Table 1. The ultimate and proximate analysis of the biomass feeds and their bio-oils (as received).

Quantity	Poplar wood		Poultry litter	
	biomass	bio-oil	biomass	bio-oil
Carbon, %	46.11	57.80	35.33	62.84
Hydrogen, %	6.44	6.80	5.40	8.31
Nitrogen, %	<0.5	<0.5	4.10	7.23
Sulfur, %	<0.05	<0.05	0.69	<0.9
Oxygen, % (by diff.)	35.36	34.85	54.48	20.72
Moisture, %	6.86	3.79	8.37	3.99
Ash, %	0.94	<0.9	22.78	<5
HHV, MJ/kg	20.00	24.56	14.26	27.98

GPC curves (wf/dLog mw vs Log mw) of poplar wood and poultry litter biooils are presented in Figure 1. and Figure 2. analysis data showed that weight average molecular weights of the poplar and poultry bio-oils were 527 and 420 Da. Number average molecular weights 230 and 286 Da respectively. The polydispersities for poplar and poultry biooils were 2.289 and 1.470 and 1.26.

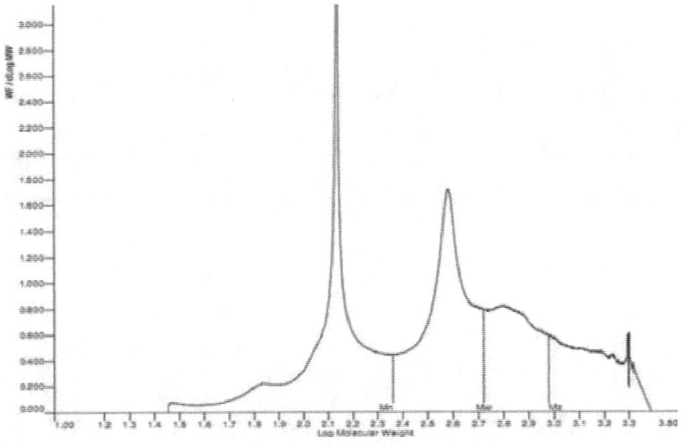

Figure 1. GPC curve of the poplar wood biooil.

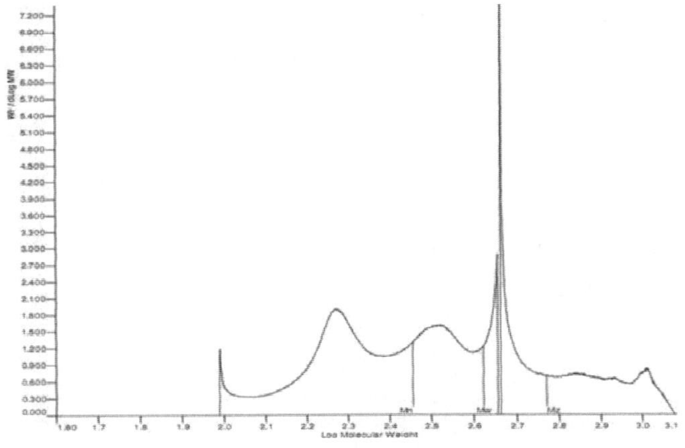

Figure 2. GPC curve of the poultry litter biooil.

The ^{13}C NMR spectras are shown in Figure 3.-4. and Table 2 presents the integrated ^{13}C spectra of the bio-oils. The poplar wood bio-oil had lower amount of carbonyl carbon than that of pine and oak wood bio-oil (3.0, 11.8, 18.1 % respectively). The poplar wood bio-oil carbohydrate carbon content (13.8%) was similar to the oak wood bio-oil (16.1%) and higher than that of the pine wood bio-oil(5.8%). The poplar wood bio-oil exhibited more aromatic (49.6:13.0), more carbohydrate (13.2:0.2) and more methoxy/hydroxyl (13.8:6.8) carbon than the poultry litter bio-oil. Remarkably, the poultry litter bio-oil exhibited more alkyl carbons (76.5%) than the poplar wood bio-oil (20.4%). The higher alkyl content of poultry litter bio-oil can be attributed to presence of proteins in the feed.

Table 2. ^{13}C NMR Integrations poplar wood and poultry litter bio-oils.

		Carbon distribution, as % of total C			
Type of carbon	Chem. shift (ppm)	Poplar wood bio-oil	Poultry litter bio-oil	Pine wood bio-oil	Oak wood bio-oil
Carbonyl	215-163	3.0	3.5	11.8	18.1
Total aromatic	163-110	49.6	13.0	48.4	40.1
General aromatic	163-125	29.4	3.8	-	-
Aromatic (guaiacyl)	125-112	18.8	9.0	-	-
Aromatic (syringil)	112-100	13.9	1.4	-	-
carbohydrate	110-84	13.2	0.2	5.8	10.3
Methoxy/hydroxy	84-54	13.8	6.8	16.1	16.1
Alkyl carbons	54-1	20.4	76.5	17.9	15.5

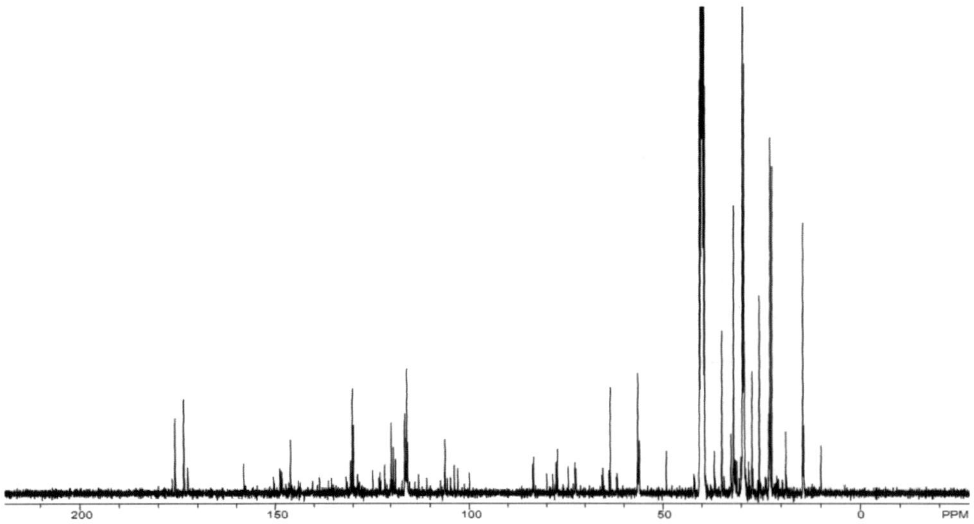

Figure 3. ^{13}C spectra of the poultry litter bio-oil.

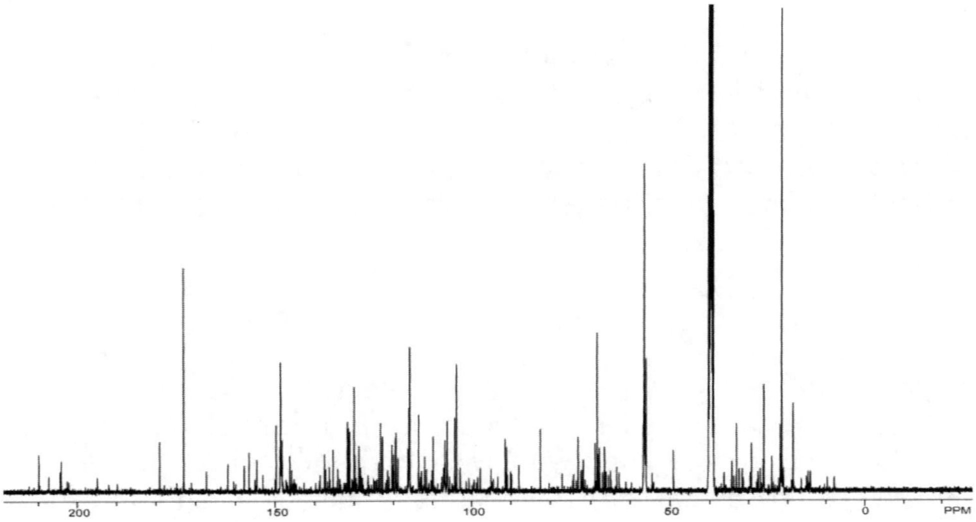

Figure 4. ^{13}C spectra of the poultry litter bio-oil.

The preliminary data showed that poplar wood and poultry litter can be utilized as a valuable biomass resource for the production biofuels and chemicals. Elemental analysis, ^{13}C NMR and GC data showed that fast pyrolysis process is an efficient tool in production of three main products from one feed in one step within seconds. Biooils produced had reasonable calorific value nearly half of the fuel oil and contained mainly aromatic and aliphatic groups that can be further processed into valuable chemicals. Utilization of poultry litter as biofuel feedstock will ease the landfill related environmental problems.

References:

1. Mohan, D.; Pittman, C. U.; Steele, P. Pyrolysis of wood/biomass- a critical review. *Energy Fuels* **2006**, *20*, 848–889.

2. Bridgwater, A. V. Renewable fuels and chemicals by thermal processing of biomass. *Chem. Eng. J.* **2003**, *91*, 87–102.

3. Bridgwater, A. V.; Peacocke, G. V. C. Fast pyrolysis for biomass. *Renewable Sustainable Energy Rev.* **2000**, *4* (1), 1–73.

4. Agblevor, F.A.; S. Besler and A.E. Wiselogel, Fast pyrolysis of stored biomass feedstocks, *Energy and Fuels* **1995**, 9, 635–640.

5. Ingram,L.; Mohan,D.; Bricka, M.; Steele, P.; *et al.*, Pyrolysis of Wood and Bark in an Auger Reactor: Physical Properties and Chemical Analysis of the Produced Bio-oils. *Energy Fuels,* **2008**, *22* (1), 614–625

SYNGAS/ HYDROGEN PRODUCTION VIA GASIFICATION OF MEAT AND BONE MEAL

Chirayu G. Soni, University of Saskatchewan, Saskatoon, SK, Canada
Zhiguo Wang, University of Saskatchewan, Saskatoon, SK, Canada
Ajay K. Dalai, University of Saskatchewan, Saskatoon, SK, Canada
Todd Pugsley, University of Saskatchewan, Saskatoon, SK, Canada
Terrance Fonstad, University of Saskatchewan, Saskatoon, SK, Canada

Introduction

Meat and bone meal (MBM) is a by-product of rendering industries. It is obtained after removal of fat from mammal carcasses during the cooking process followed by drying and crushing. [1]. MBM has historically been used worldwide in the formulation of feed for cattle because of its high protein content, but after being found to be responsible for the spreading of bovine spongiform encephthalopathy (BSE), its use has become progressively restricted [2]. MBM is produced in large amount and requires a safe disposal method to avoid the spreading of BSE. The responsible prions for BSE pathogens can be destroyed by combustion at 850 °C for at least 2 sec, or at 3 bar and 133 °C for at least 20 min [3]. In most of European Union countries, MBM is either incinerated or burned in cement kilns. The high temperatures (1300-1500 °C) inside cement kiln combined with sufficient residence times destroy the responsible prions [4].

MBM has a high calorific value of approximately 20.6 MJ/kg (Dry basis). Therefore, the other thermal treatment processes of pyrolysis, gasification and combustion represent alternatives to incineration that may be used to destroy the responsible prions for BSE while recovering the fuel value of the MBM in liquid or gaseous form. Gasification is one of the thermochemical processes by which a carbonaceous feedstock such as biomass or coal can be converted into a gaseous energy carrier by partial oxidation at elevated temperature. The product gases mainly contain CO, H_2, CO_2, CH_4 and other hydrocarbons [5]. They can be used to power gas engines or gas turbines, or can be catalytically converted into methanol, dimethylether, Fischer-Tropsch oils or other chemical products [6, 7, 8]. Research has been conducted on pyrolysis, co-combustion and gasification of MBM. Ayllon et al. [3] studied the effect of final pyrolysis temperature (300-900 °C) and heating rate (2, 8 and 14 °C /min) on the fixed bed pyrolysis of MBM. They observed that final pyrolysis temperature is more important than the heating rate. Chaala et al. [9] studied vacuum pyrolysis of MBM at a temperature of 500 °C with total pressure of 4 kPa and heating rate of 15 °C /min. They found pyrolytic oil and solid residue as main products with small amount of gases. Fedorowicz et al. [10] studied the gasification of feed consisting of MBM, cow carcasses and two types of SRMs at 1000 °C with steam and nitrogen as carrier gases in bench scale gasification system. A major issue encountered was the production and handling of tar. Other studies include fluidized bed co-combustion of MBM with coals and olive bagasse [11], co-combustion of coal and MBM in fluidized bed [12] and the use of AB fluidized bed combustor for the utilization of raw animal wastes [13].

The aim of the present work is to study the effects of process parameters on hydrogen/syngas production via gasification of MBM followed by thermal cracking of tar in two-stage fixed bed reaction system.

Experimental procedure

Gasification of MBM followed by thermal cracking of tar was carried out at atmospheric pressure in a two-stage fixed bed reactor system. The first stage was used for gasification of MBM while the second stage was used for further cracking of tar. The experimental set up is shown in Fig. 1. The first stage and second stage reactors were made up of Inconel tubing having 10.5 mm ID and 500 mm and 300 mm lengths respectively. The feed material was placed inside the first stage reactor and inert packed bed material (Ottawa sand) was placed inside the second stage reactor. These material were supported on the plug of quartz wool which was supported on the supporting mesh inside each reactor.

Fig. 1. Schematic diagram of experimental setup for gasification of meat and bone meal in two-stage fixed bed reaction system

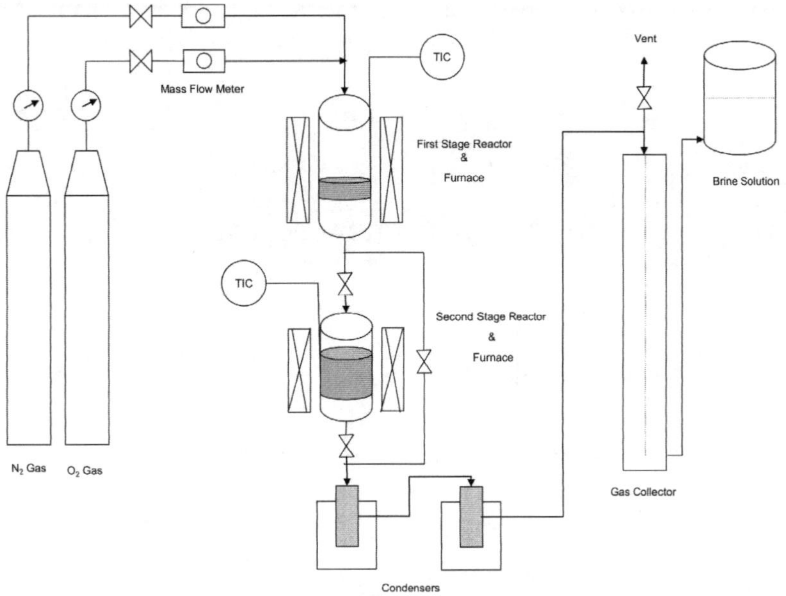

The sample size was kept 2 g for all experiments. Both the reactors were connected by a connector of 3 mm diameter tube 40 mm length and placed inside separate tubular furnaces. K-type of thermocouples were directly connected with each reactor to maintain and monitor the temperature of the reactor via temperature controller system (Eurotherm models 2132 and 2416, USA). Nitrogen was used as an inert carrier gas while pure oxygen was used as a source of oxidant in the reaction. Both gases were supplied at the desired flow rate from separate cylinders through needle valve and mass flow meter (Aalborg model GFM17). Two glass condensers in series at the bottom of the second stage reactor, surrounded by mixture of ice and salt, were used to condense the tar and cool down the product gases. The product gases were collected in saturated brine solution column and the amount of gases collected was measured by the displacement of the brine solution. The saturated brine solution was used to prevent CO_2 dissolution in pure water. The heating rate of the first stage reactor was kept at 25 °C /min. The oxygen supply was started when the first stage reactor temperature reached to 50 °C. It took 24 to 32 min. from 50 °C to reach the final temperature of 650 to 850 °C in the case of single stage experiments. In the case of two-stage experiments, the second stage reactor was heated to the desired temperature before the heating of first stage started. The heating of first stage was started at 25 °C /min. The reminder of the experimental procedure was the same as single stage experiments. After attaining the final desired temperature of the first stage, the reaction was allowed to continue for the next 30 min. Subsequently the heating was stopped and the reactor(s) was allowed to cool down. The amount of product gases collected was measured and analyzed using two different gas chromatographs (GCs HP 5880 and HP 5890). The amount of condensed tar in the glass condensers and the char left inside the reactor were measured. After each run, the reactor and glass condensers were cleaned and dried for the next run.

Results and discussion

Effect of temperature during single stage operation

Temperature is a very important parameter in gasification processes. The effect of final temperature of the first stage was studied by simply bypassing the second stage reactor of Fig. 1. In present work, the final temperature of first stage was varied from 650 to 850 °C with an increment of 50 °C while holding equivalence ratio (ER) at 0.2 and nitrogen flow rate at 45 ml/min. The effect of final temperature of the first stage on product (char, tar and gas) yield gas composition is presented in Fig.2. As expected, char (from 22.93 to 17.78 wt. %) and tar (from 47.03 to 40.71 wt. %) yields decreased,

Fig. 2. Effect of final temperature of first stage on product yield and gas composition at ER of 0.2 and N_2 flow rate of 45 ml/min

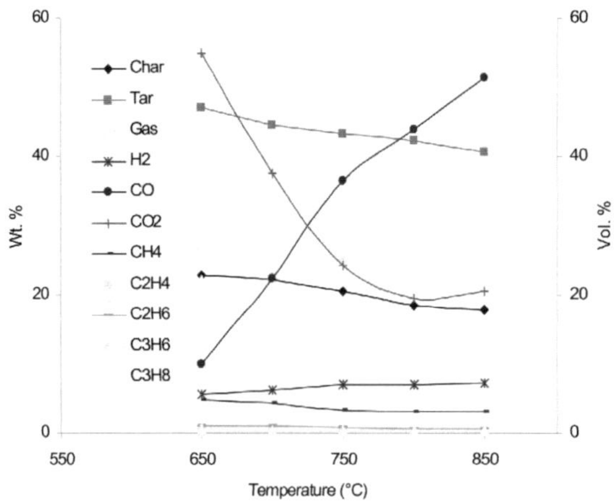

whereas gas yield increased (from 22.13 to 30.83 wt. %) with an increase in temperature from 650 to 850 °C. Higher temperature is favorable for gasification of char as well as thermal cracking of tar. Similar trends were also observed by other researchers [14, 15, 16]. The product gas was mainly composed of H_2, CO, CO_2, CH_4 and other heavier hydrocarbon gases. The H_2 yield increased from 5.52 to 7.26 vol. % (N_2 free basis). The CO_2 yield decreased sharply from 55.02 to 20.5 vol. % and the CO yield increased remarkably from 9.97 to 51.57 vol. % with a temperature increase from 650 to 850 °C. More carbon conversion and Boudard reaction, reaction of carbon and carbon dioxide to produce carbon monoxide, at higher temperature are partially responsible for the observed increase in CO and the decrease in CO_2 content. CH_4 and other hydrocarbons showed slight variation with increase in temperature. This could be explained by the likelihood of some secondary reactions with CO_2.

Effect of second stage temperature during two-stage operation

The tar can be removed by many ways. Increase in residence time and thermal cracking are two of them [5]. The second reactor stage was introduced in series to the first stage for further cracking of tar and to increase the H_2 yield. It was packed with an inert material, Ottawa sand. The second stage reactor temperature was varied from 650 to 850 °C with an increment of 50 °C keeping temperature of first stage at 850 °C, ER at 0.2, nitrogen flow rate at 45 ml/min and packed bed height at 60 mm. The effect of temperature of second stage on product yield and gas composition is presented in Fig. 3. The tar yield decreased from 26.48 to 17.4 wt. % and gas yield increased from 42.3 to 52.17 wt. % as second stage temperature increased from 650 to 850 °C.

Fig. 3. Effect of second stage temperature on product yield and gas composition at first stage final temperature of 850 °C, ER of 0.2, N_2 flow rate of 45 ml/min and packed bed height of 60 mm

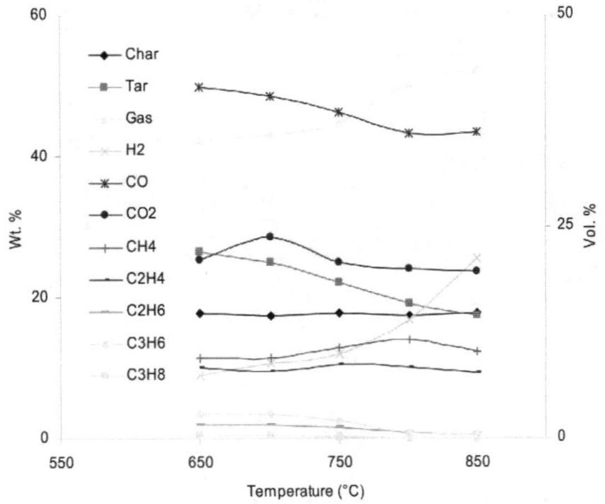

Up to 750 °C, the decrease in tar yield and increase in gas yield were 16.42 and 5.58 % but after 750 °C, they were 21.37 and 16.82 % respectively. This shows that thermal cracking of tar is highly favorable after 750 °C. After introducing the second stage, the tar yield is 57.25 % lower while gas yield is 40.9 % greater than that obtained using only a single stage reactor configuration. CO content varied little from 41.57 to 36.2 vol. %, while that of CO_2 first increased slightly from 21 to 23.8 vol. % and then decreased up to 19.7 vol. %. The CH_4 yield increased after 700 °C up to 800 °C, while other heavier hydrocarbons decreased sharply after 700 °C. This phenomenon can be explained by cracking of heavier hydrocarbons into lighter ones and H_2 which can be further supported by rapid increase in H_2 content after 700-750 °C. Similar kind trends were obtained by other researchers [18, 19].

Effect of equivalence ratio

The equivalence ratio is a second key parameter in gasification processes using air/oxygen. In the present work, ER was varied from 0.15 to 0.3 with an increment of 0.05, keeping final temperature of first stage and second stage temperature at 850 °C and the packed bed height at 60 mm. The nitrogen flow rate was varied accordingly to compensate for the variation in oxygen flow rate. The effect of ER on product distribution and gas composition is shown in Fig. 4. It was observed that with increase in ER, char and tar yield decreased from 20 to 14.23 wt. % and 19.58 to 14.6 wt.% respectively while gas yield increased from 43.33 to 56.23 wt.%. The increase in ER increases oxidation reactions and deteriorates the product gas quality. It was observed that H_2 content reached a maximum value of 21.2 vol. % at ER of 0.2 and then dropped to 16.92 vol. % with further increase in ER. An increase in CO_2 content from 16.8 vol. %

Fig. 4. Effect of equivalence ratio (ER) on product yield and gas composition at first stage final and second stage temperatures of 850 °C and packed bed height of 60 mm

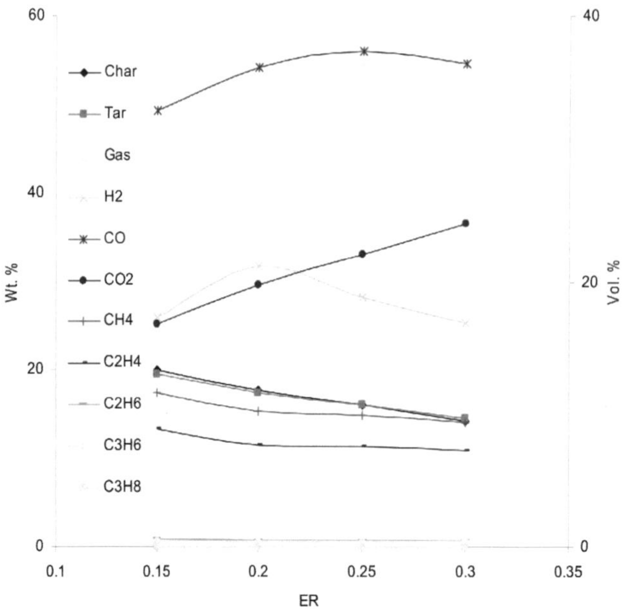

to 24.44 vol. % with increase in ER can be explained by strong oxidation reactions. The CH_4, C_2H_4 and C_2H_6 content in the product gas were decreased marginally from 11.6 to 9.4, 8.9 to 7.26 and 0.61 to 0.51 vol. % respectively. There was no significant deviation observed in percentages of higher hydrocarbon gases. Similar trends were observed by other researchers [20, 21, 22].

Effect of packed bed height

The residence time can be altered by changing the packed bed height of the inert material in cracking (second) stage. Higher residence time is favorable for cracking of tar, higher hydrocarbons as well as secondary reactions to take place. In the present work, the packed bed height of the cracking stage was varied from 40 to 100 mm while keeping final temperature of first stage and second stage temperature at 850 °C, ER at 0.2 and N_2 flow rate at 45 ml/min. It was found that an increase in packed bed height decreased tar yield from 18.58 to 14.23 wt.% and increased gas yield from 50.6 to 54.55 wt.%. The H_2 content increased little from 18.86 to 22.25 vol. % with packed bed height. The CO content initially decreased from 39.3 to 36.2 vol. % and then increased up to 40.65 vol. %. There was a little variation observed in case of CO_2 content. It was almost constant initially and then decreased slightly from 19.7 to 19.17 vol. %. In the case of CH_4, it decreased slightly from 10.7 to 9.81 vol. % and then increased slightly up to 10.33 vol. %. Other heavier hydrocarbons showed the opposite trend than CH_4. They first increased and then decreased. An initial increase in heavier hydrocarbons could be because of cracking of tar into heavier hydrocarbon gases and the ensuing decrease is likely due to further cracking into CH_4 and H_2. Xiao et al. [22] studied the effect of bed height in air gasification of polypropylene plastic waste in fluidized bed gasifier. They found that tar and heavier hydrocarbons yield decreased while gas yield increased with increase in bed height.

Conclusions

MBM can be effectively used as an alternate source of gaseous fuel as well as for production of hydrogen/syngas. The higher temperature of 850 °C of both the stages was found to be favorable for higher hydrogen/syngas production within the range of temperature studied. In comparison to the single stage process, the two- stage process proved to be effective to increase the H_2 yield from 7.26 to 22.25 vol. % and to increase the gas yield from 30.83 to 54.55 wt. %. At the same time, the tar yield was reduced from 40.71 to 14.23 wt. %. The equivalence ratio of 0.2 was found to be optimum to yield maximum H_2 production of 21.2 vol. % at the final temperature of first stage and second stage temperature of 850 °C, N_2 flow rate of 45 ml/min and packed bed height of second stage at 60 mm. The effect of packed bed height had a less significant effect. The tar yield decreased a little from 18.58 to 14.23 wt. % whereas that of gas increased from 50.6 to 54.55 wt. %. There was no significant variation observed in gas compositions.

Acknowledgement

The authors are grateful to the University of Saskatchewan, Canada for the financial support for this project.

References

1. Bradley R. Eur J Epidemiol 1991; 7:532-44.
2. Rodehutscord M, Abel HJ, Friedt W, Wenk C, Flachowsky G, Ahlgrimm HJ, Johnke B, Kuhl R, Breves G. Arch. Anim. Nutr. 2002; 56: 67-91.
3. Ayllon M, Aznar M, Sanchez JL, Gea G, Arauzo J. Chemical Engineering Journal 2006; 121: 85-96.
4. Conesa JA, Fullana A, Font R. Journal of Anal. Appl. Pyrolysis 2003; 70: 619-630.
5. Bridgwater AV. Fuel 1995; 74:631-53.
6. McKendry P. Bioresource Technology 2002; 83: 55-63.
7. Han J, Kim H. Renewable & Sustainable Energy Reviews 2006; doi: 10.1016/j.rser.2006.07.015.
8. Bridgwater T. J Sci Food Agric 2006; 86: 1755-68.
9. Chaala A and Roy C. Environ. Sci. Technol. 2003; 37: 4517-22.
10. Fedorowicz ME, Miller SF, Miller BG. Energy & Fuels 2007; 21: 3225-32.
11. Fryda L, Panopoulos K, Vourliotis P, Pavlidou E, Kakaras E, Fuel 2006; 85: 1685-99.
12. Gulyurtlu I, Boavida D, Abelha P, Lopes MH, Cabrita I, Fuel 2005; 84: 2137-48.
13. Pilawska M, Baron J, Zukowski W, Kandefer S. The Seventh Asia-Pacific International Symposium on Combustion and Energy Utilization, Hong Kong SAR, proceedings 15-17 December 2004.
14. Franco C, Pinto F, Gulyurtlu I, Cabrita I. Fuel 2003; 82: 835-42.

15. Wei L, Xu S, Zhang L, Liu C, Zhu H, Liu S. International Journal of Hydrogen Energy 2007; 32: 24-31.
16. Lv PM, Xiong ZH, Chang J, Wu CZ, Chen Y, Zhu JX. Bioresource Technology 2004; 95: 95-101.
17. Devi L, Ptasinski KJ, Janssen FJJG, Van Paasen SVB, Bergman PCA, Kiel JHA. Renewable Energy 2005; 30: 565-587.
18. Beck SR and Wang MJ. Ind. Eng. Chem. Process Des. Dev. 1980; 19: 312-317.
19. Turn S, Kinoshita C, Zhang Z, Ishimura D, Zhou J. Int. J. Hydrogen Energy 1998; 23 (8): 641-648.
20. Manya JJ, Sanchez JL, Abrego J, Gonzalo A, Arauzo J. Fuel 2006; 85: 2027-2033.
21. Mansaray KG, Ghaly AE, Al-Taweel AM, Hamdullahpur F, Ugursal VI. Biomass and Bioenergy 1999; 17: 315-332.
22. Xiao R, Jin B, Zhou H, Zhong Z, Zhang M. Energy Conversion & Management 2007; 48: 778-786.

97e Intermediate Temperature Catalytic Reforming of Bio-Oil for Distributed Hydrogen Production

Jonathan R. Marda[1], Anthony M. Dean[1], Stefan Czernik[2], Robert J. Evans[3], Richard French[2], and Matthew Ratcliff[2]. (1) Chemical Engineering, Colorado School of Mines, Alderson Hall, 1600 Illinois Street, Golden, CO 80401, (2) National Renewable Energy Laboratory, Golden, CO 80401, (3) Hydrogen Technologies & Systems Center, National Renewable Energy Laboratory, Golden, CO 80401-3393

Introduction

With the world's energy demands rapidly increasing, it is necessary to look to sources other than fossil fuels, preferably those that minimize greenhouse emissions. One such renewable source of energy is biomass, which has the added advantage of being a near-term source of hydrogen. While there are several potential routes to produce hydrogen from biomass thermally, given the near-term technical barriers to hydrogen storage and delivery, distributed technologies such that hydrogen is produced at or near the point of use are attractive. One such route is to first produce bio-oil via fast pyrolysis of biomass close to its source to create a higher energy-density product, then ship this bio-oil to its point of use where it can be reformed to hydrogen and carbon dioxide. This route is especially well suited for smaller-scale reforming plants located at hydrogen distribution sites such as filling stations. There is also the potential for automated operation of the conversion system.[2]

A system has been developed for volatilizing bio-oil with manageable carbon deposits using ultrasonic atomization and by modifying bio-oil properties, such as viscosity, by blending or reacting bio-oil with methanol. Non-catalytic partial oxidation of bio-oil is then used to achieve significant conversion to CO with minimal aromatic hydrocarbon formation by keeping the temperature at 650°C or less and oxygen levels low. The non-catalytic reactions occur primarily in the gas phase. However, some nonvolatile components of bio-oil present as aerosols may react heterogeneously. The product gas is passed over a packed bed of precious metal catalyst where further reforming as well as water gas shift reactions are accomplished completing the conversion to hydrogen.[4]

The approach described above requires significantly lower catalyst loadings than conventional catalytic steam reforming due to the significant conversion in the non-catalytic step. The goal is to reform and selectively oxidize the bio-oil and catalyze the water gas shift reaction without catalyzing methanation or oxidation of CO and H_2, thus attaining equilibrium levels of H_2, CO, H_2O, and CO_2 at the exit of the catalyst bed. Experimental

Bio-oil (mixed with varied amounts of methanol to reduce the viscosity and homogenize the bio-oil) or selected bio-oil components (such as methanol or HAA) are introduced at a measured flow rate through the top of a vertical quartz reactor which is heated using a five zone furnace. The ultrasonic nozzle used to feed the reactants allows the bio-oil to flow down the center of the reactor at a low, steady flow rate. Additionally, the fine mist created by the nozzle allows for intimate mixing with oxygen and efficient heat transfer, providing optimal conditions to achieve high conversion at relatively low temperatures in the non-catalytic step thus reducing the required catalyst loading. Generation of the fine mist is especially important for providing good contact between non-volatile bio-oil components and oxygen. Oxygen and helium are also delivered at the top of the reactor via mass flow meters with the amount of oxygen being varied to maximize the yields of H_2 and CO and the amount of helium being adjusted such that the gas phase residence time in the hot zone is ~0.3 and ~0.45 s for bio-oil and methanol experiments, respectively.

A catalyst bed can be located at the bottom of the reactor tube. To date, catalyst screening experiments have used Engelhard noble metal catalysts. The catalysts used for these experiments were 0.5 % rhodium, ruthenium, platinum, and palladium (all supported on alumina). Experiments were performed using pure alumina as well. Both the catalyst type and the effect of oxygen and steam on the residual hydrocarbons and accumulated carbon containing particulates were investigated. The residence time before the catalyst is varied to determine the importance of the non-catalytic step and its potential effect on the required catalyst loading. Non-catalytic experiments (primarily homogeneous cracking) use a bed of quartz placed to capture any deposits that are formed in the volatilization and cracking zones.

The inner reactor effluent is quenched by a flow of 10 SLPM He which serves to sweep the products quickly (~0.03 s) to a triple quadrupole molecular beam mass spectrometer (MBMS) for analysis. The MBMS serves as a universal detector and allows for real time data collection. The study of pyrolysis by MBMS has been described previously.[5] The dilution of the reactor effluent reduces the potential problems caused by matrix effects associated with the MBMS analysis. Argon is used as an internal standard in the quantitative analysis of all the major products (CO, CO_2, H_2, H_2O, and benzene) as well as any residual carbon, which is determined by subsequent oxidation of carbon (monitored as CO_2) after shutting off the feed and maintaining the oxygen/helium flow. Any carbon left on the walls of the reactor (measured by weighing) is also counted as residual carbon.

Results and Discussion

Non-Catalytic Partial Oxidation. To date, experiments have been performed using 50:50 and 70:30 bio-oil:methanol (wt. %) mixtures at 650°C. Earlier experiments showed that temperatures above 600°C yielded the most promising results. However, it should be noted that a significant amount of bio-oil conversion occurs at temperatures as low as 550°C.

The maximum CO yields from the bio-oil (expressed as percentage of carbon converted to CO after the CO attributed to methanol is subtracted out) for the 50:50 mixture (~50 %) and the 70:30 mixture (~55 %) were remarkably similar and occurred at nearly the same effective molar oxygen to carbon (O:C) ratio (~1.5). The effective O:C ratio does not include the oxygen due to the water content of the bio-oil is subtracted out (since water is an inert under these conditions).

The maximum overall H_2 yields calculated as percent of the amount of hydrogen that could be produced by stoichiometric conversion of bio-oil to CO_2 and H_2 (the H_2 attributed to methanol is included) for the 50:50 (~23 %) and 70:30 (~24 %) mixtures are nearly identical and occur at an effective O:C ratio of ~1.5. It should be noted that the maximum H_2 and CO yields occurred at the same O:C ratios.

Catalytic Auto-thermal Reforming.Limited catalyst screening has identified rhodium supported on alumina as the most promising catalyst since it yields results (~84% yield for H_2 and ~61% yield for CO) that come closest to equilibrium predictions of 80% H_2 yield and 69% CO yield (see Table 1). The other catalysts used fell well short of equilibrium (28 to 34% H_2 yield and CO yield ranging from 45 to 51%). However, the use of any of the catalysts does result in a significant increase in the H_2 and CO (13% H_2 yield and 39% CO yield). Methanol conversion also increases significantly from 49% in the non-catalytic experiments to 100% (the value predicted by equilibrium) for the rhodium catalyst. The other catalysts yielded methanol conversions ranging from 75 to 88%. Hence, the catalyst serves a dual purpose of increasing the conversion of the components of the bio-oil as well as catalyzing the water gas shift reaction.

It should be noted that the catalysts used were chosen because they were on hand and no efforts have been made to date to optimize the catalyst size or loading. Future experiments will involve exploring the optimal size and loading of the catalyst as well as determining how effective the non-catalytic partial oxidative step is at reducing the catalyst loading. A suitable catalyst that will minimize or eliminate the use of expensive rhodium is highly desirable and efforts will be made to discover such a catalyst.

The non-catalytic results described above will be talked about briefly as these results will be discussed in great detail in a separate presentation. The focus of this presentation will be on the catalytic results.

Table 1. Preliminary catalytic screening results for 50:50 bio-oil:methanol mixture at an oxygen to carbon ratio of 1.3. Product selectivities are given as weight percent of the initial carbon (or hydrogen) entering the system that is converted to a particular species.

	CH_3OH % conversion	CO wt. % C	CO_2 wt. % C	H_2 wt.% H	H_2O wt. % H
Equil.	100	69	24	80	13
Non-catalytic	49	39	7	13	38
Alumina	39	36	7	9	38
Pd	82	51	10	34	41
Pt	88	45	12	31	45
Rh	100	61	23	84	18
Ru	75	48	9	28	41

References

(1) Milbrandt A., *A Geographic Perspective on the Current Biomass Resource Availability in the United States*; National Renewable Energy Laboratory: Golden, CO, TP-560-39181, **2005**.

(2) Czernik, S.; Elam, C.; Evans R.; and Milne, T., in *Thermal and Chemical Biomass Conversion*; Bridgwater, A.V. and Boocock D.G.B., eds.; CPL Press: Newbury, UK, 2006; pp.1752-1761.

(3) U.S. Department of Energy: Hydrogen, Fuel Cells and Infrastructure Technologies Program; *Multi-Year Research, Development and Demonstration Plan, Section 3.1 Hydrogen Production*; U.S. Department of Energy: Office of Energy Efficiency and Renewable Energy: Washington, D.C, 2006.

(4) Evans, R.J.; Czernik, S.; French, R.; and Marda, J.; *Distributed Bio-Oil Reforming, DOE Hydrogen Program FY2006 Annual Progress Report*, 2006.

(5) Evans, R. J.; and Milne, T. A., *Energy & Fuels*, 1, **1987**, 123-137.

97f Biomass Gasification Using a Fixed-Bed Downdraft Gasifier

Morteza Eslamian[1], Alice Yip[1], **Masahiro Kawaji**[2], C.A. Mims[1], and Honghi N. Tran[3]. (1) Chemical Engineering and Applied Chemistry, University of Toronto, 200 College Street, Toronto, ON M5S 3E5, Canada, (2) Department of Chemical Engineering and Applied Chemistry, University of Toronto, 200 College Street, Toronto, ON M5S 3E5, Canada, (3) Pulp & Paper Centre, Dept. of Chemical Engineering & Applied Chemistry, University of Toronto, 200 College Street, Toronto, ON M5S 3E5, Canada

BACKGROUND

The lime kiln, being an important unit in kraft pulping process, consumes a lot of energy in the form of natural gas or crude oil. The annual cost of the production of 1000 ADt/d of CaO from lime kilns is more than $6 million. As the cost of energy continues to increase, it is essential to find alternative fuel sources that are both suitable and economical to use in kraft pulp mills. Biomass, which refers to living and recently dead biological material, such as vegetation, or agricultural waste, is one of alternative fuel sources. It can be converted into a synthesis gas and possibly be used as a fuel in the lime kiln and other processes. This can potentially reduce the energy cost as well as net CO2 emission.

Biomass gasification has been widely studied over the years to convert wood waste and forestry residues into a combustible synthesis gas (syngas). In gasification, up to 80% of the energy in the biomass is converted into the gaseous form [1]. The produced syngas has the potential to be used in a wide range of systems to generate heat and electricity.

A gasification process is the partial combustion of biomass in the reactor with a gasifying agent such as steam, air, or pure oxygen at a temperature of about 800-1000oC [2]. The produced syngas is composed of H2, CO, CO2 and a small quantity of CH4. The heating value of the syngas depends on the gasification medium and the operating conditions, but roughly it could be about one-third that of the natural gas (37MJ/Nm3) [1]. It is known that the composition of the syngas produced from biomass gasification depends not only on the feedstock, but also on the gasifier design as well as gasification conditions [3].

The gasifier designs may be grouped into one of three categories: updraft [4], downdraft [5, 6] and fluidized bed gasifiers [7]. Due to the unique nature of the gasifier design, the characteristics of the produced syngas vary dependent on the gasifiers used, making the results hardly comparable. It is thus advantageous to have a single gasifier to systematically compare the characteristics of the produced syngas obtained from a broad range of the biomass of interest, making the results readily available to the industry. The feedstock under study includes wood waste, saw dust, slash, mill sludge, peat and agricultural residues, etc.

The approach of this research is to design and construct a bench-scale fixed-bed gasifier with biomass processing capacity of several kilograms per hour and to develop a standard characterization procedure. Then, the effect of the operation conditions, such as the air flow rate and the process temperature on the characteristics of the syngas will be studied.

EXPERIMENTAL

A downdraft gasifier has been designed and constructed for this study. It consists of a stainless steel outer cylinder, which is 56 cm high, 28 cm in diameter, and has a wall thickness of 8 mm. The combustion occurs within a smaller circular cylinder placed inside the outer cylinder. This reaction chamber is made of high temperature steel and is 15.2 cm high, 12.7 cm in diameter and 4 mm in wall thickness.

A grate is attached to the bottom of the reaction chamber to hold the feedstock. This grate is shaken regularly by two rods to let the ash fall on the base plate of the main chamber. Several feed-through holes have been devised on the base plate to let the oxidant (air) and thermocouples enter the chamber. Six type K thermocouples (Omega) are inserted into the reaction chamber from the base plate to continuously monitor and record the process temperature, 1 cm away from the reaction chamber inner wall and also on the chamber centerline.

Two 150 W cartridge heaters are inserted and secured on the reaction chamber walls. These heaters initiate the combustion of their surrounding biomass, once they are switched on for a short duration. Air is used as the gasification medium in this study. Four ¼ inch OD tubes uniformly supply air to four locations of the reaction chamber wall. A pressure regulator and a rotameter control the air flow rate from the compressed air line to the chamber.

The feedstock is fed to the chamber from the top. A funnel directs the feedstock to the reaction chamber. During the burning process, a lid is placed to close the chamber. Therefore the syngas has to move downward through the grate holes to enter the annulus between the reaction chamber and outer cylinder. The syngas then exits the chamber via two tubes welded on the chamber wall, and is directed to a Bunsen burner positioned in a fume hood where the syngas burns. A second Bunsen burner connected to a natural gas line is used to facilitate the burning of the synthesis gas. A valve is mounted on the exit syngas pipe line for sampling. For sampling this valve opens and the sample synthesis gas is taken using a gas tight syringe. A thermocouple mounted on the exit line measures the gas temperature at the exit.

PRELIMINARY EXPERIMENTS

Operating the gasifier in steady state conditions, i.e. having a stable burning of the biomass, while the process temperature is nearly constant, is one of the most important issues in gasification. Preliminary experiments revealed that when the combustion was initiated without preheating the chamber and the biomass, a larger amount of tar was produced which was condensed in the form of biofuel at the bottom of the outer chamber. This is because of the low temperature of the process, which hinders the cracking and reduction of large molecules of tars to light-weight hydrocarbons such as CH4. To resolve this problem, the chamber was preheated by the purge nitrogen gas. However, due to sever heat loss in the pipelines, no significant rise in the biomass and reaction chamber temperature was observed.

As a second way to preheat the chamber, the 150 Watt cartridge heater was contiguously turned on and off at 15 s intervals to warm up the chamber. This was done, once the chamber was purged by nitrogen gas and before the gasification air was introduced to the chamber.

The temperature profiles of the reaction chamber centerline and wall, and the syngas during the warm-up period showed that the temperature profiles level off after about 2200 s. This indicates that just using the 150 Watts cartridge heater, the maximum temperature achievable is about 300°C.

During the gasification process including the warm-up period for an air injection rate of about 10 L/min, the temperature within the reaction chamber decreased first upon introduction of air, and then started to increase once the combustion was initiated. The maximum temperature of about 800°C was recorded. To investigate whether the combustion could be sustained without keeping the heater on, the heater was turned off, however, the rapid temperature drop indicated that the combustion was terminated once the heater was turned off. Therefore, the heater was turned on again and the temperature increased again due to overheating of the heater. The heater suffered a failure and the temperature started to decrease.

It was observed that a continuous flow of white smoke was produced and exited the gas line. Also, a dilute oily liquid was condensed on the bottom of the outer chamber. Preheating the chamber resulted in the formation of a lower amount of tar, which was also less concentrated compared to the case where the chamber was not preheated.

FUTURE PLAN

The gasifier has been constructed and successfully tested with wood chips. The first goal is to achieve steady-state operation of the apparatus and also to determine the burning rate and therefore the valid range of the air flow rate that could be employed. Then, for a valid range of air flow rate and temperature below 1,000°C, the syngas will be characterized to determine its composition and also the tar content.

REFERENCES

[1] Austermann, S., Whiting, K. J., "Commercial Assessment: Advanced Conversion Technology (Gasification) for Biomass Projects", Biomass Energy Centre, UK (2007).

[2] U.S. Department of Energy, "Biomass Program: Thermochemical Platform." Energy Efficiency and Renewable Energy, (2007).

[3] Paisley, M. A., Biomass Energy. "Kirk Othmer Encyclopedia of Chemical Technology". John Wiley & Sons, Inc. June 2007.

[4] Yang, W., Ponzio, A., Lucas, C., and Blasiak, W. "Performance Analysis of a Fixed-Bed Biomass Gasifier Using High-Temperature Air", Fuel Processing Technology Vol. 87, pp. 235-245 (2006).

[5] Lv, P., Yuan, Z., Ma, L., Wu, C., Chen, Y., Zhu, J. "Hydrogen-rich Gas Production from Biomass Air and Oxygen/steam Gasification in a Downdraft Gasifier." Renewable Energy Vol. 32, pp. 2173-2185, (2007).

[6] Garcia-Bacaicoa, P., Bilbao, R., Arauzo, J., and Salvador, M. L. "Scale-Up of Downdraft Moving Bed Gasifiers (25-300kg/H)- Design, Experimental Aspects and Results", Bioresource Technology Vol. 48, pp. 229-235 (1994).

[7] Li, X. T., Grace, J. R., Lim, C. J. ,Watkinson, A. P., Chen, H. P. and Kim, J. R. "Biomass Gasification in a Circulating Fluidized Bed", Biomass and Bioenergy Vol. 26, pp. 171-193, (2004).

TB Topical B: Emerging Energy: Frontiers in Research and Innovation

Bioethanol Production by Fermentation of Starch Particles using a Novel, Inexpensive Media

Nancy Bawa, Gordon A. Hill and Catherine Niu
Department of Chemical Engineering, University of Saskatchewan
Saskatoon, Saskatchewan, Canada

INTRODUCTION

An efficient method for conversion of biomass into fuel is by ethanol production because ethanol is an economical as well as environmentally friendly liquid fuel. Ethanol has the advantages of being renewable, cleaner burning and producing no net greenhouse gases (1). Yeast cells (*Saccharomyces cerevisiae*) are facultative anaerobes and under anaerobic conditions can ferment glucose to ethanol. *Saccharomyces* yeasts are ideal for ethanol production due to several properties including fast growth rates, efficient glucose repression, efficient ethanol production and a tolerance for environmental stresses, such as high ethanol concentration and low oxygen levels. Glucose is broken down to form pyruvate in most organisms via the glycolytic pathway and this pyruvate can result in ethanol under anaerobic conditions. The energy for growth of cells during ethanol production is provided by the glycolytic and fermentation pathways (2). But glucose is an expensive raw material for industrial applications and therefore, starch is an economical alternative.

Although sugarcane is the most widely used raw material for ethanol production via fermentation, starch based raw materials are more abundant. In the USA and Eastern Canada, corn is used for ethanol production, but in Western Canada, wheat starch is more readily available. Approximately 25 million tones of wheat are grown every year in Canada, 50% of which is produced in Saskatchewan. The starch content of this wheat is approximately 65% based on a dry mass basis. Starch, a macromolecular polymer of glucose units, is a significant component of domestic and commercial waste and a useful resource that can be converted into ethanol. Yeast cells are unable to consume raw starch directly and hence, the starch must first be broken down into simple sugars (3). This task is achieved by two enzymes: alpha amylase, which hydrolyzes alpha-1, 6 linkages and glucoamylase, which hydrolyzes alpha-1, 4 linkages in starch molecules. In industrial processes, starch is first cooked at high temperatures to solubilize it, followed by high temperature enzymatic liquefaction (α-amylase) and saccharification (glucoamylase). The use of starch particles for bioethanol production by fermentation is economical because it avoids this cooking process and saves time and energy. However, developing a cost-effective fermentation process for starch particles with high efficiency of ethanol production still remains an outstanding challenge.

A significant aspect in the fermentation of biomass to ethanol is the cost of the medium used. Kadam *et al.* in 1997 used a low-cost medium containing 0.3% corn steep liquor and 2.5 mM $MgSO_4.7H_2O$ (4). In this work, simultaneous cold starch hydrolysis and fermentation was investigated as a way of ethanol production from pure wheat starch. With a view to minimize process costs while maintaining a satisfactory ethanol yield, a novel, inexpensive media was formulated and tested for its fermentation capacity. This novel media, like the media used by Kadam *et al.*, consists of commercially available components and this makes the process relevant at an industrial level. The new media formulation and bioreactor operating conditions are shown to significantly improve the economics of bioethanol production from raw starches.

MATERIALS AND METHODS

Microorganism: Commercially available Muntons Active Brewing Yeast was used for this project.

Substrates and Enzymes: Two substrates were used for the batch runs, glucose (VWR) and starch particles (Archer Daniels Mdlnd). The α-amylase and glucoamylase enzymes were purchased from Valley Research and the barley malt was purchased from InfraReady Products Limited.

Preparation of media: The expensive medium used was similar to that of Thatipamala, Rohani and Hill (5). A 100 g/L glucose media solution contained (per liter of distilled water): 100 g of glucose; 10 g of yeast extract (technical grade, Difco laboratories); 2.5 g of ammonium chloride; 2.91 g of disodium hydrogen phosphate; 3.0 g of potassium dihydrogen phosphate; 0.25 g of magnesium sulfate; 0.08 g of calcium chloride; 4.3 g of citric acid and 3.0 g of sodium citrate. For 20 g/L glucose media solution, 3.0 g of yeast extract was used. The amounts of nutrients (excluding the buffer components) were increased proportionately when preparing 200 g/L glucose media solution. The pH during growth remained between 4.0 and 4.2. The nutrient solution with yeast extract and the glucose were sterilized separately.

The above mentioned expensive media was used for batch runs at different substrate concentrations and the results were compared with those of the same batch runs using a novel, inexpensive media. This new media contains cheaper ingredients as compared to the expensive media that contains pure chemical compounds. For the starch particle runs, the expensive media contained Valley Research alpha amylase and glucoamylase and the novel, inexpensive media contained ground barley malt and glucoamylase. Correct proportions of the novel media components of the inexpensive media were used based on the substrate concentration.

Experimental Setup: A New Brunswick Scientific Bioflo III bioreactor was used in this study. The working volume in the bioreactor was 2.0 L and 100 mL of inoculum was added for each run. Batch fermentation runs using three different substrate concentrations of 20 g/L, 100 g/L and 200 g/L were performed. The temperature was varied to improve productivity.

During the initial hydrolysis phase, the bioreactor was operated at four different temperatures, 30^0C, 37.5^0C, 45^0C and 52.5^0C and the agitation was set to 375 RPM. The air flow rate was set to 0.02 L/minute. Three mL samples were collected at various time intervals as the yeast cells multiplied. To obtain biomass concentrations, the optical density was measured for every sample collected and the remaining samples were filtered using 0.2 μm cellulose nitrate membrane filters (Whatman) and stored for ethanol and glucose analyses. For the starch particle runs, a 1 mL sample was collected and diluted using 10 mL media and 50 μL of this diluted sample was used for cell counts using a Petrov-Hauser counting grid. Also, for the starch particle runs performed using the novel, inexpensive media, 3 mL samples were collected, vacuum filtered and the solids collected on the 0.45 μm cellulosic filter paper were dried in a vacuum oven overnight at 65^0C.

Measurement of Biomass, Glucose, Starch and Ethanol Concentrations: The optical densities were measured using a UV-VIS mini 1240 spectrophotometer (Shimadzu) at 620 nm wavelength. A graph of optical density verses dry weight was linear up to an optical density of 2.5 with a slope of 1.51 g/L. Glucose concentrations were measured using a glucose analyzer, GM8 Analyser (Analox Instruments) based on the glucose oxidase principle. Ethanol concentrations were measured using gas chromatography using a capillary column with a flame ionization detector (FID). The initial and final starch concentrations were measured using the iodine method (6) for the runs performed using the expensive media.

RESULTS AND DISCUSSION

Production of Ethanol from Glucose using Muntons Active Brewing Yeast:

An ethanol-prolific, plasmid-free yeast cell line i.e. Muntons active brewing yeast was cultured in the bioreactor at glucose concentrations of 20, 100 and 200 g/L. Figures 1 and 2 below show the production of biomass, utilization of glucose and production of ethanol with time during the culturing of Muntons yeast cells in traditional, expensive 100 g/L glucose media and in novel, inexpensive 100 g/L glucose media, respectively.

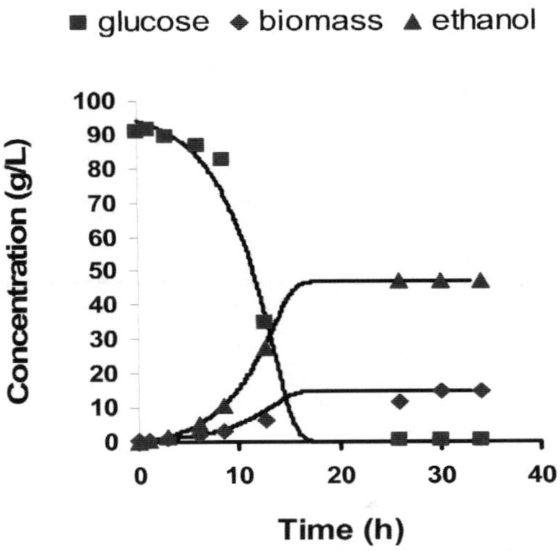

Figure 1: Glucose utilization and biomass and ethanol production curves for batch experiments in expensive media containing 100 g/L glucose

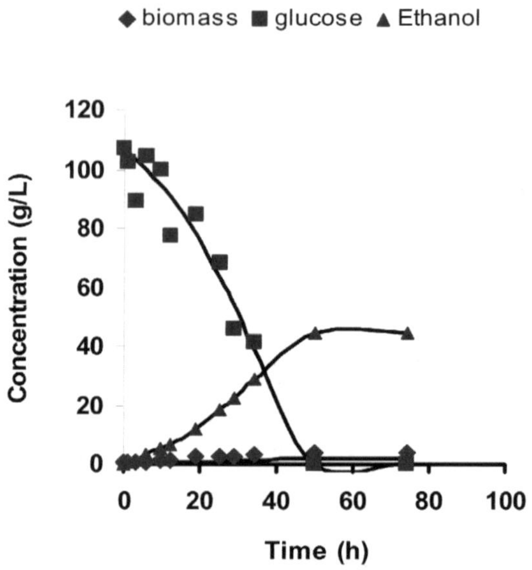

Figure 2: Glucose utilization and biomass and ethanol production curves for batch experiments in novel, inexpensive media containing 100 g/L glucose

Table 1 below shows a summary of the maximum specific growth rates (μ) and the ethanol yields for the batch runs performed with glucose as substrate.

Table 1: Summary of glucose batch runs

	EXPENSIVE	**INEXPENSIVE**
Glucose conc. (g/L)	20	20
μ (h^{-1})	0.32	0.12
Ethanol Yield	0.45	0.44
Glucose conc. (g/L)	100	100
μ (h^{-1})	0.084	0.090
Ethanol Yield	0.50	0.42
Glucose conc. (g/L)	200	200
μ (h^{-1})	0.13	0.11
Ethanol Yield	0.51	0.70

From the above summary it can be concluded that the performance of the novel, cheap media is as good as the expensive media. The ethanol yields are comparable and hence, this novel media was used for the starch particle batch runs.

Production of Ethanol from Starch Particles using Muntons Active Brewing Yeast:

Muntons active brewing yeast was cultured in the bioreactor at starch particle concentrations of 20, 100 and 200 g/L in both expensive and inexpensive media. These batch runs were performed at 30^0C. Three additional runs were performed using the novel media containing 100 g/L starch particles with hydrolysis carried out at 37.5^0C, 45.0^0C and 52.5^0C, respectively.

Table 2 below shows a summary of the maximum specific growth rates (μ) and the ethanol yields for the batch runs performed with starch particles as substrate.

Table 2: Summary of starch particle batch runs

	EXPENSIVE (30^0C)	INEXPENSIVE (30^0C)	INEXPENSIVE (37.5^0C)	INEXPENSIVE (45^0C)	INEXPENSIVE (52.5^0C)
Starch particles (g/L)	20	20	-	-	-
μ (h^{-1})	0.25	0.16	-	-	-
Ethanol Yield	0.41	0.46	-	-	-
Starch particles (g/L)	100	100	100	100	100
μ (h^{-1})	0.13	0.11	0.10	0.10	0.13
Ethanol Yield	0.34	0.44	0.37	0.44	0.39
Starch particles (g/L)	200	200	-	-	-
μ (h^{-1})	0.12	0.068	-	-	-
Ethanol Yield	0.42	0.43	-	-	-

As seen from the above table, the novel, inexpensive media runs with starch particles as substrate gave comparable results to those achieved with expensive media. The ethanol yields are comparable to those obtained from the expensive media runs.

Hence, it can be concluded that the novel, inexpensive media is as good as the expensive media and can be used for ethanol production with high yields. Moreover, the use of starch particles as substrate avoids the high temperature cooking procedure, thereby, making this entire fermentation process for bioethanol production more efficient and economical than current technology used in the bioethanol industry in North America.

REFERENCES

1. Altıntaş, M., Ülgen, K. Ö., Kırdar, B., Önsan, Z. I. and Oliver, S. G. (2002) Improvement of ethanol production from starch by recombinant yeast through manipulation of environmental factors. *Enzyme and Microbial Technology* **31**, 640-647.
2. Pisˇkur, J., Rozpe˛dowska, E., Polakova1, S., Merico, A. and Compagno, C. (2006) How did *Saccharomyces* evolve to become a good brewer? *TRENDS in Genetics* **22** (4), 183-186.
3. Birol, G., Önsan, Z. İ., Kirdar, B. and Oliver, S. G. (1998) Ethanol production and fermentation characteristics of recombinant *Saccharomyces cerevisiae* strains grown on starch. *Enzyme and Microbial Technology* **22**, 672-677.
4. Kadam, K. L. and Newman, M. M. (1997) Development of a low-cost fermentation medium for ethanol production from biomass. *Appl Microbiol Biotechnol* **47**, 625-629.
5. Thatipamala, R., Rohani, S and Hill, G. A. (1992) Effects of High Product and substrate Inhibitions on the Kinetics and Biomass and Product Yields during Ethanol Batch Fermentation. *Biotechnology and Bioengineering* **40**, 289-297.
6. Xiao, Z., Storms, R. and Tsang, A. (2006) A quantitative starch-iodine method for measuring alpha-amylase and glucoamylase activities. *Analytical Biochemistry* **351**, 146-148.

173b Simulation of Simultaneous Sacharification and Fermentation of Ethanol Production from Cellulose by An Operable Model

Jiacheng Shen, biological systems engineering, virginia tech, sertz hall, blacksburg, VA 24061 and Foster A. Agblevor, Bilogical Systems Engineering, Virginia Polytechnic Institute and State University, Seitz hall, Blacksburg, VA 24061.

Abstract

An operable batch model of simultaneous saccharification and fermentation (SSF) for ethanol production from cellulose has been developed. The model includes four ordinary differential equations that describe the changes of cellobiose, glucose, yeast and ethanol concentration with respect to residence time. These equations were used to simulate the data for the four main components in the SSF process of ethanol production from Avicel PH101. The model parameters at 95% confidence intervals were determined by a MATLAB program based on the batch experimental data of the SSF without additional experiments. The model was used to show that the influence of the initial parametric conditions on the ethanol production as follows: initial cellulose concentration > initial enzyme concentration > initial yeast concentration. The model also showed that the conversion of cellulose to cellobiose was the rate-controlling step in a series reaction of cellulose to ethanol. The rate constants of cellobiose to glucose and glucose to ethanol were 5.3 and 5.6 times that of cellulose to cellobiose, respectively. The batch model was extended to the continuous and fed-batch operating models. For the continuous operation in the SSF, the productivities increased with increasing dilution rate, until a maximum value was attained, and rapidly decreased as the dilution rate approached the washout point. The model also predicted a relatively high ethanol mass for the fed-batch operation than the batch operation.

Keywords: Simultaneous saccharification and fermentation; Ethanol; Cellulose; Model; Operating mode

173c Synthesis of Liquid Fuel from Woody Biomass by A Bench-Scale BTL Plant

Toshiaki Hanaoka, Yanyong Liu, Satoshi Hirata, and Kinya Sakanishi. Biomass Technology Research Center, National Institute of Advanced Industrial Science and Technology, Suehiro 2-2-2, Hiro, Kure, Hiroshima, 737-0197, Japan

Transportable liquid fuel such as hydrocarbon and dimethyl ether (DME) produced from biomass as a sustainable energy resource is an extremely promising carbon neutral fuel because CO2 generated by the use of liquid fuel produced from biomass can be offset by CO2 fixed by photosynthesis. In addition, this liquid fuel is clean, emitting none of the air pollutants such as SOx, NOx, and particulate matter. In particular, hydrocarbon fuel such as gasoline, light oil, and kerosene is applicable to the conventional engine. Accordingly, in view of petroleum saving and environmental protection, the synthesis process of hydrocarbon fuel as a transportable liquid fuel from biomass is one of the highest interests among the conversion technologies of biomass to energy. In the present work, on a bench-scale, the conversion of biomass into a transportable liquid fuel (BTL) system consisting of biomass gasification, gas cleaning, and FT synthesis was constructed. The synthesis of hydrocarbon liquid fuel from woody biomass was sequentially carried out. Eucalyptus tip (30 mm squares) was employed as a woody biomass sample. The ultimate analysis (daf basis) was as follows; C: 51.0 wt%, H: 6.0 wt%, N: 0.2 wt%, S: 0.015 wt%, O (diff.): 42.8 wt%. The proximate analysis was as follows; moisture: 10.5 wt%, volatile matter: 73.1 wt%, fixed carbon: 15.8 wt%, ash: 0.6 wt%. This bench-scale BTL plant consists of gasifier, scrubber, compressor, desulphurization tower, decarbondioxide installation, and FT synthesis reactor. A downdraft fixed bed gasifier was employed in the oxygen-enrichment gasification step. The biomass feed rate was approximately 9 kg/h on a dry basis. The oxygen concentration in the gasifying agent was in the range of 21.0 to 31.5 vol.%. The temperature of the combustion zone and reduction zone inside the gasifier was from 900 to 1000 oC, from 800 to 850 oC, respectively. The product gas flow rate was adjusted 18 Nm3/h using a roots blower installed downstream of the gasifier. The sulfur compounds derived from woody biomass gasification deactivate the FT catalyst. Therefore, in the gas cleaning step, the concentration of sulfur compounds has to be decreased less than 1 ppm. In this process, the sulfur compounds such as H2S and COS were removed by a scrubber and desulphurization tower packed with activated carbon. In the gas compression step, the cleaned gas has to be compressed up to several MPa in order to enhance liquid fuel yield in the FT synthesis step. Moreover, the partial pressure of syngas was increased by removing CO2 from cleaned gas on the decarbondioxide installation. In the FT synthesis step, a slurry bed reactor was employed. The inner diameter was 100 mm, and the length was 1500 mm. Hexadecane was employed as a solvent. 600 g of Co catalyst was used. Prior to run, the catalyst was reduced for 10 hours at 400 oC. The reaction pressure and temperature were 3 MPa and in the range of 280 to 300 oC. The slurry was stirred at 650 rpm. The feed gas rate was 18.5 NL/min. The reaction time was for 6 hours. The obtained gas and liquid fuel were quantitatively analyzed by GC-TCD, GC-FID, and GC-FPD. With increasing the oxygen concentration in the gasifying agent, the carbon conversion on a carbon basis increased from 91.9 to 96.2 C-mol%, and CO increased from 22.8 to 30.1 vol.%, and H2 concentration increased from 16.8 to 23.7 vol.%. CO2 and CH4 concentrations were approximately 11 and 2 vol.% and independent of the oxygen concentration. In the oxygen concentration of 27.6 vol.%, the product gas was 21.9 vol.% of H2, 29.3 vol.% of CO, 11.0 vol.% of CO2, 2.0 vol.% of CH4, 0.4 vol.% of C2 hydrocarbon, 35.5 vol.% of N2, 9.1 ppm of H2S, and 5.3 ppm of COS . The H2S, COS, and CO2 in the obtained gas in this case were removed through the desulphurization tower and decarbondioxide installation during gas compression up to 13 MPa. As a result, the cleaned gas was 25.2 vol.% of H2, 30.8 vol.% of CO, 40.6 vol.% of N2, 2.5 vol.% of CH4 0.9 vol.% of CO2, less than 5 ppbv of H2S, and less than 5 ppbv of COS. The hydrocarbon liquid fuel was produced from the cleaned gas by the FT reactor. As a result, the CO conversion was 36.2%, H2 conversation was 76.7%. The selectivity of CO2, CH4, C2, C3, C4, and C5+ was 14.6%, 8.3%, 1.3%, 1.6%, 1.2%, and 73.0%, respectively. From this result, 1.9 L of liquid fuel was obtained from 13 kg of woody biomass on a dry basis by the bench-scale BTL plant per day. In the future work, we are planning to scale up to the production capacity of 16 L of liquid fuel per day.

173d Effect of the Anion of Ionic Liquid for Dissolution of Cellulose

Hyunjoo Lee[1], Eun Hee Jeon[2], Dinh Quan Nguyen[1], Byoung Sung Ahn[1], and Hoon Sik Kim[2]. (1) Energy & Environment Research Division, Korea Institute of Science & Technology, 39-1, Hawolgok-dong, Seongbuk-gu, Seoul, South Korea, (2) Department of Chemistry, Kyung Hee University, 1 Hoegi-dong, Dongdaemoon-gu, Seoul 130-701, South Korea

Cellulose, the most abundant polymer from nature, is regarded as an important source for bio-ethanol and bio-butanol. In this experiment, cellulose dissolution and regeneration have been conducted by using of 1-butyl-3-methylimidazolium-based ionic liquids having different anion. It was found that the degree of dissolution was closely related with coordinating ability of the anion in ionic liquid, that is, less coordinating or non-coordinating anions such as BF4-, PF6- and TFSI- showed poor solubility for cellulose, while coordinating anions like halide, carboxylate, bicarbonate dissolve cellulose up to 20 wt % depending on the experimental condition. After dissolution, the cellulose was regenerated and then analyzed by TGA, SEM, IR, and XRD. Depending on the heating method and ionic liquid used, regenerated celluloses showed different morphology and chemical properties.

Effect of flow characteristics and temperature on performance and efficiency of a direct methanol fuel cell stack

Liang QI1), Rong LI1), Shu-bo WANG1), Jin-hai WANG1), Xiao-feng XIE1,*), V. K. MATHUR2)

1. Institute of Nuclear and New Energy Technology, Tsinghua University, Beijing 100084, China

2. Department of Chemical Engineering, University of New Hampshire, NH 03824, USA

Corresponding author Tel: +86 10 6278 4827; fax: +86 10 6277 1150
Email address: xiexf@tsinghua.edu.cn (Xiao-feng Xie)

Abstract
The paper focuses on several factors influencing the performance and efficiency of a 40-cell direct methanol fuel cell (DMFC) stack for portable power applications. The factors include methanol flow rate, air flow rate and methanol inlet temperature. In particular, the dependence of Faraday efficiency and overall energy efficiency on these operating conditions is analyzed under constant power output. Although under optimum conditions power, Faraday efficiency and overall energy efficiency are improved to some extent, the methanol crossover would always cause considerable efficiency loss if Nafion electrolyte membrane is used. Highly conductive membranes with significantly lower methanol crossover are necessary for the successful application of the DMFC stacks. Apart from optimum operating conditions, homogeneous distribution of reactants in the stack and other inherent characteristics of individual cells are equally important for the stack to provide stable and effective performance.

Keywords: Direct methanol fuel cell stack (DMFC stack); Performance; Faraday efficiency; Overall energy efficiency; Flow rate; Temperature

1. Introduction

Fuel cells are considered to be the best substitute for conventional power source, because of multiple advantages such as high energy conversion efficiency, broad range of available fuels, and near zero pollutant emissions. When using high energy density methanol as a fuel directly, the fuel cells called the direct methanol fuel cells (DMFCs) are being developed to provide power to electrical facilities, small and middle scale power stations as well as for portable electrical applications [1-10].

However, there are still several barriers limiting the performance of DMFCs. First, the anode electro-oxidation kinetics is poor which causes high anode overpotential and consequently decreases the overall cell voltage. It is believed that the slow oxidation kinetic of methanol to carbon dioxide results in the formation of carbon monoxide as an intermediate which is strongly absorbed on the P_t catalyst surface [2]. Second, the anode methanol solution penetrates through the polymer electrolyte membrane to the cathode and reacts directly with the oxidant at the cathode catalyst layer. This methanol crossover not only brings about a waste of the fuel, but also a mixed overpotential which decreases the overall cell voltage. In addition, the performance of a DMFC is strongly influenced by several other factors under actual operation.

The most common influencing factors are the DMFC operating temperature, methanol solution concentration and the fuel or oxidant flow rates. Research workers [11-13] have discussed how the cell performance changes with the operating temperature in the range of 30°C to 120°C and methanol concentration changing from 0.5M to 4M. Ge et al. [14] studied the effect of cell operating conditions such as temperature, fuel flow rate, air flow rate as well as cathode gas humidification. It is found that all operating parameters, except the cathode gas humidification, had significant effect on DMFC cell performance. Kim et al. [15] investigated the characteristics of a 50W DMFC stack under various operating parameters. They found that a maximum power of 54W could be achieved with the supply of a 2M methanol solution. Liu et al. [16] investigated methanol crossover, water transfer and power characteristics of a 40W DMFC stack under various experimental conditions including different methanol concentrations, flow rates at anode and cathode and flow directions of reactant fluids. The counter-flow arrangement of reactants produced 20% higher power output than by co-flow arrangement. Buttin et al. [17] studied the performance of a single cell and a 150W DMFC stack assembled with the similar single cells. The lower performance of a DMFC stack compared to a singe cell was mainly attributed to the non-uniform distribution of reactants and the higher internal resistance in the DMFC stack. Dohle et al. [18] studied a 500W DMFC stack and reported system performance and optimal operating conditions. Gogel [19] investigated the effects of operating parameters and electrode structures on the performance and methanol permeation of a 12-cell DMFC stack.

Chu et al. [20, 21] studied energy conversion efficiency and Faraday efficiency for a passive DMFC. The performance and overall efficiency of a DMFC using non-fluorinated composite membranes were evaluated by Silva et al. [2]. The effect of operating temperature on cell performance and overall efficiency was discussed. Liu et al. [22] investigated the effect of membrane thickness and fuel utilization in passive DMFC using different Nafion membranes.

In this paper, a DMFC stack consisting of 40 single cells with an active area of 27.5 cm^2 per cell has been developed and the effect of various operating conditions on Faraday efficiency and overall energy efficiency under constant power has been investigated.

2. Energy efficiency of a DMFC stack

2.1 Theoretical thermodynamic efficiency

In a DMFC, the following catalytic reactions take place:

anode reaction: $CH_3OH(l) + H_2O(l) \xrightarrow{Pt/Ru} CO_2(g) + 6H^+ + 6e$

cathode reaction: $3/2\,O_2(g) + 6H^+ + 6e^- \xrightarrow{Pt} 3H_2O(l)$

overall reaction: $CH_3OH(l) + 3/2O_2(g) \rightarrow CO_2(g) + 2H_2O(l)$

Through these electrochemical reactions, the chemical energy of fuel methanol is converted directly into electric energy. According to the second law of thermodynamics:

$$\Delta G = \Delta H - T\Delta S \quad (1)$$

The theoretical open circuit voltage, U_{th} of a DMFC is calculated from the Gibbs free energy, ΔG as given below where n is the number of the electrons transferred during the reaction and *F* is the Faraday constant.

$$U_{th} = \frac{-\Delta G}{nF} \quad (2)$$

Where n=6 and *F*=96485 c/mol.

The theoretical thermodynamic efficiency of a fuel cell can be defined as:

$$\eta_{th} = \frac{-\Delta G}{-\Delta H} \quad (3)$$

2.2 Voltage efficiency

The cell reaction is irreversible and the cell potential during operation is affected by Ohmic losses, overvoltages at the electrodes as well as the occurrence of a mixed potential at the cathode. During cell operation the cell voltage, U_C is determined as:

$$U_C = U_{th} - \eta_a - \eta_c - \eta_{ac} - \eta_R \quad (4)$$

Where η_a, η_c, η_{ac} and η_R are the anode overvoltage, cathode overvoltages, mixed potential at the cathode and Ohmic losses, respectively. The lower kinetics of the DMFC anode is responsible for substantially higher overvoltages (approximately 200-300mV) in comparison to the hydrogen anode overpotentials for proton exchange membrane fuel cells (PEMFCs). The mixed potential, η_{ac} lowers the cell voltage in the range of 50-200mV depending on operating conditions such as methanol concentration and operating temperature [23]. Moreover, the conductivities of the polymer electrolyte membrane and other components in DMFCs are finite. This accounts for another major potential loss. The Ohmic losses, η_R also decreases the cell voltage. The open circuit voltage, U_0 of a DMFC is lower than the theoretical value, U_{th} and the cell voltage U_C is even lower than the open circuit voltage U_0. Therefore, we need to define the voltage efficiency η_U to evaluate the cell voltage performance:

$$\eta_U = \frac{U_C}{U_{th}} \tag{5}$$

2.3 Fuel efficiency, Faraday efficiency, and Current efficiency

Mass diffusion between anode and cathode reduces the open circuit potential and the performance of chemical power sources. In DMFCs, methanol diffusion or crossover from the anode to the cathode lowers fuel utilization, increases cathode polarization and causes excess thermal load in the cell and consequently lowers the cell performance.

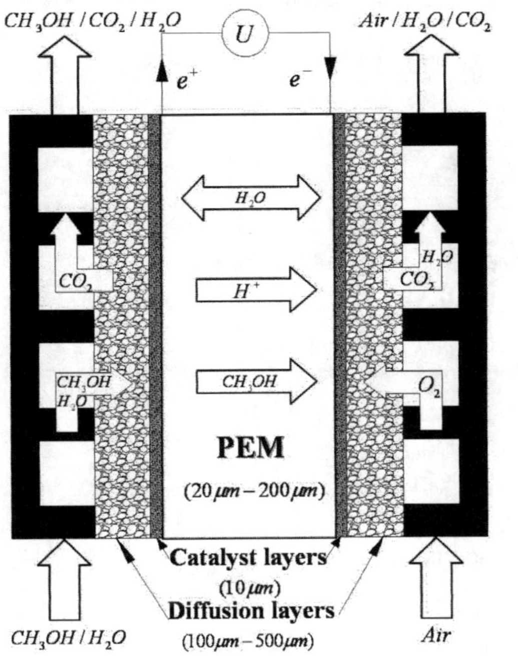

Fig. 1 Schematic of a DMFC single cell

Fig.1 shows the basic process in a DMFC. The anode inlet methanol solution $N^{anode}_{CH_3OH,in}$ is consumed due to several processes. Part of the total inlet methanol, expressed as $N^{anode}_{CH_3OH,ele}$ is electrochemically converts into electric current at the anode. The oxidation product of methanol at anode is regarded as only CO_2 produced at high surface area of PtRu catalyst. Secondly, the unreacted methanol, $N^{anode}_{CH_3OH,out}$ leaves the cell together with anode reaction precuts. The third portion of the methanol, $N^{perm}_{CH_3OH}$ permeates from the anode to the cathode. Most of it is oxidized at cathode to produce CO_2 and H_2O. This methanol crossover is merely a waste of fuel as it does not provide any electric energy, yet raises issues such as the presence of a mixed overpotential which decreases the cell voltage.

There are two causes for methanol crossover from the anode to the cathode: the methanol diffusion from anode to cathode and the methanol transport by electro-osmosis. The cause of the methanol diffusion is the concentration gradient across the membrane which is maintained by the oxidation of methanol reaching the cathode. The amount of solvated methanol molecules transported by electro-osmosis is mainly a function of the current density, the methanol concentration at the anode-membrane interface and the membrane properties. During cell operation, the overall methanol balance can be expressed as:

$$N^{\text{anode}}_{\text{CH}_3\text{OH,in}} = N^{\text{anode}}_{\text{CH}_3\text{OH,ele}} + N^{\text{perm}}_{\text{CH}_3\text{OH}} + N^{\text{anode}}_{\text{CH}_3\text{OH,out}} \tag{6}$$

Total methanol consumption, N_{CH_3OH} is given by:

$$N_{\text{CH}_3\text{OH}} = N^{\text{anode}}_{\text{CH}_3\text{OH,in}} - N^{\text{anode}}_{\text{CH}_3\text{OH,out}} = N^{\text{anode}}_{\text{CH}_3\text{OH,ele}} + N^{\text{perm}}_{\text{CH}_3\text{OH}} \tag{7}$$

The fuel efficiency, η_M can be defined as the ratio of methanol flow, $N^{anode}_{CH_3OH,ele}$, which generates electric current to the overall methanol consumption, N_{CH_3OH}:

$$\eta_M = \frac{N^{anode}_{CH_3OH,ele}}{N_{CH_3OH}} \tag{8}$$

Assuming that every cell in the stack has the same characteristics (actually, cells in a stack perform distinctively from one another), we can have the following equation:

$$N^{i}_{CH_3OH,j} = z\int_0^t n^{i}_{CH_3OH,j}dt \tag{9}$$

where i and j represent the different parts of methanol flow in the cell during operation and z is the number of single cells in a stack. For a single DMFC: $z = 1$. Equations (2.6) and (2.7) can be rewritten to give the total methanol, n_{CH_3OH} (mol/s) consumption as :

$$n^{anode}_{CH_3OH,in} = n^{anode}_{CH_3OH,ele} + n^{perm}_{CH_3OH} + n^{anode}_{CH_3OH,out} \tag{10}$$

and

$$n_{CH_3OH} = n^{anode}_{CH_3OH,in} - n^{anode}_{CH_3OH,out} = n^{anode}_{CH_3OH,ele} + n^{perm}_{CH_3OH} \tag{11}$$

According to the Faraday's law, $n^{anode}_{CH_3OH,ele}$ can be calculated from the electric current, I_{ele}:

$$n^{anode}_{CH_3OH,ele} = \frac{I_{ele}}{6F} \tag{12}$$

Thus, using Eqs. (2.8) and (2.11), we can describe the fuel efficiency, η_M as:

$$\eta_{\mathrm{M}} = \frac{\int_0^t n^{\mathrm{anode}}_{\mathrm{CH_3OH,ele}}dt}{\int_0^t n_{\mathrm{CH_3OH}}dt} = \frac{\int_0^t I_{\mathrm{ele}}dt}{6F\int_0^t n_{\mathrm{CH_3OH}}dt} \tag{13}$$

In some studies [2, 20, 21], Faraday efficiency, η_F is introduced to examine how the fuel is converted in a single cell or a stack. Faraday efficiency is defined as a ratio of actual capacity to theoretical capacity of fuel consumed in a fuel cell:

$$\eta_F = \frac{z\int_0^t idt}{6FN_{CH_3OH}} \tag{14}$$

Comparing Eq. (13) and Eq.(14), it can be concluded that estimating actual energy conversion from the fuel chemical energy, the fuel efficiency and the Faraday efficiency are the same.

According to the Faraday's law, $n^{anode}_{CH_3OH,ele}$ can be calculated from the electric current, I_{ele} as shown in Eq.(2.12). Similarly, permeation methanol, $n^{perm}_{CH_3OH}$ can be written in terms of parasitic current, I_{perm}:

$$n^{perm}_{CH_3OH} = \frac{I_{perm}}{6F} \tag{15}$$

The current efficiency, η_I is usually defined as:

$$\eta_I = \frac{I_{ele}}{I_{ele} + I_{perm}} \tag{16}$$

Combining Eqs.(2.11), (2.12) and (2.15), the current efficiency can be calculated as:

$$\eta_I = \frac{n^{anode}_{CH_3OH,ele}}{n^{anode}_{CH_3OH,ele} + n^{perm}_{CH_3OH}} = \frac{n^{anode}_{CH_3OH,ele}}{n_{CH_3OH}} \quad (17)$$

For steady state operation of DMFC stack, the current efficiency and the fuel efficiency are the same. Anyone of the three efficiencies, namely, η_M , η_F , and η_I can be used to estimate methanol conversion to produce electric power to the overall methanol consumption.

2.4 Overall energy efficiency

In general, there are many factors that strongly influence the fuel cell efficiency. Combining Eqs. (2.3), (2.5) and (2.14), the overall energy efficiency for a single cell or DMFC stack can be calculated as: $\eta = \eta_{th} \cdot \eta_U \cdot \eta_F$

3. Experimental

The experimental DMFC stack consists of 40 cells with an active area of 27.5cm^2 per cell, two copper plates working as current collectors and two epoxy resin end plates at both ends of the stack arrangement. The cell pitch, the distance between the membrane electrode assembly (MEA), is 3.2mm. Nafion-115 (Dupont®) is used as membrane. As for experimental procedure details can be found elsewhere [16, 24].

Shown in Fig.2, the DMFC system comprises the stack, two methanol tanks, pumps, heat exchangers, steam separators, an air compressor, as well as a temperature control unit. One of the methanol tanks contains pure methanol and the other contains methanol-water solution. By mixing the flows from these two tanks, the anode inlet methanol solution is prepared and its concentration is monitored by a methanol sensor. A gas chromatograph (SHIMAZU GC-14B) with a flame ionization detector (FID) is used to measure the outlet methanol concentration. An electronic load (ITECH IT8512C) is used and experimental data are recorded by a data acquisition/switch unit (Agilent 34970A). All the data are taken when the stack operation is stable.

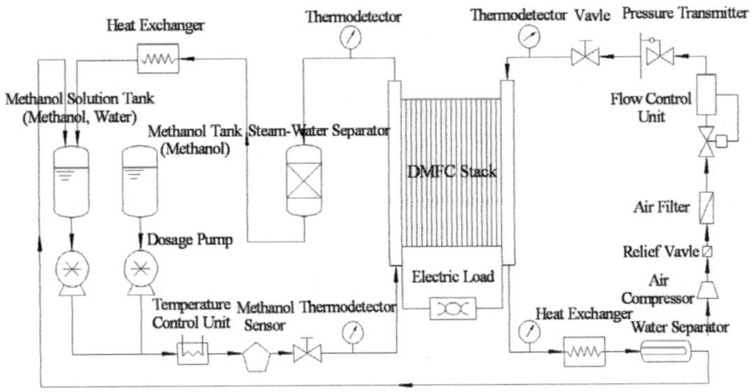

Fig. 2 A diagrammatic sketch of DMFC system

4. Results and discussion

The amount of fuel used in the stack consists of two parts. First is the electrochemically reacted fuel which contributes to generating electricity. Second is the un-electrochemically reacted fuel which contributes to the cross-over fuel. The former processes release the electrons, which are conducted from the anode to the cathode via the external circuit to generate electricity or via the interior circuit to generate heat. But at a different anode or cathode over potential, the output power is different though the amount of electrons conducted via the circuit is the same. Thus, the faradic efficiency could evaluate the efficiency of fuel utilization, which is important for the fuel cell to operate for a long time. The energy efficiency or the voltage efficiency could be used to evaluate the working capacity.

the total amount of CO_2 produced is divided into two parts. One is oxidized in the anode by the catalytic way, the other is oxidized in the cathode by the combustion way. But the part produced in the anode could diffuse to the cathode, so we can not evaluate the faradic efficiency by only calculating the amounts of CO_2 in one side. When the methanol arrives the anode catalyst layer, only part of it is converted by the electrochemistry way, which converts the chemical energy into electrical energy. From the view of point, we define the faradic efficiency as follows:

$$\eta_I = \frac{\int_0^t I_{\mathrm{ex}}dt}{z\mathrm{F}\int_0^t \Delta_{\mathrm{in}}^{\mathrm{out}} c_{\mathrm{MeOH}}dt} = \frac{n_{\mathrm{M,ca}}}{n_{\mathrm{M,ca}} + n_{\mathrm{M,ma}}} = \frac{n_{\mathrm{CO2,ca}} + n_{\mathrm{CO2,mc}}}{n_{\mathrm{CO2,ca}} + n_{\mathrm{CO2,mc}} + n_{\mathrm{CO2,cc}}} \quad (18)$$

Fuel efficiency is not the only criterion for the performance optimization. The output power which the load required must be met by the stack. Either the faradic efficiency, energy efficiency or voltage efficiency must be considered together versus fuel management, which includes methanol concentration, fuel flow rate and current.

Fig.4 illustrates the relationship between all of the parameter and efficiency (*I*=5A). The utilization of fuel is economy at the lowest MeOH concentration and the lowest flow rate (i.e. 90ml,0.5M), but the voltage efficiency attains maximum value at optimum conditions (0.45V@120ml,1.25M). When the methanol concentration is above 2 M,the faradic efficiency is lowed to 3% due to crossover. The percentage crossover methanol ranges from 7%-17%.it should also be noted that it is not possible to optimize both the faradic efficiency and voltage efficieny. The former requires low kinetic factors such as low temperatures, low fuel supply because it depends on the transport process significantly, the latter thermodynamic factors such as high temperature, high fuel supply.

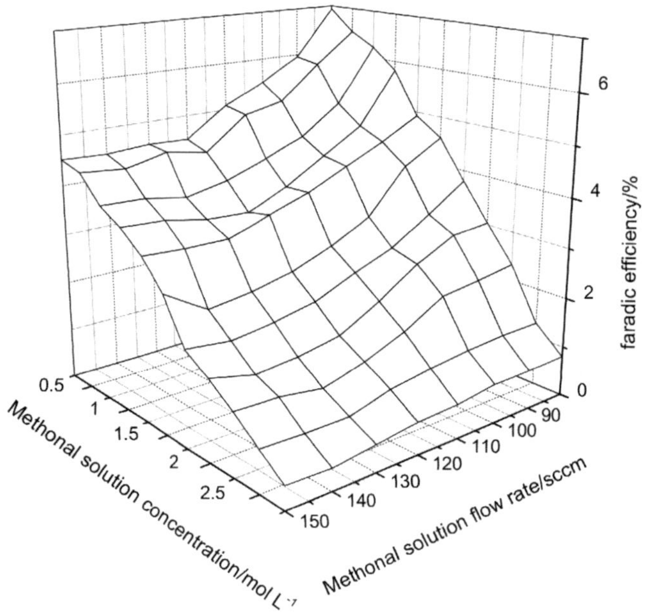

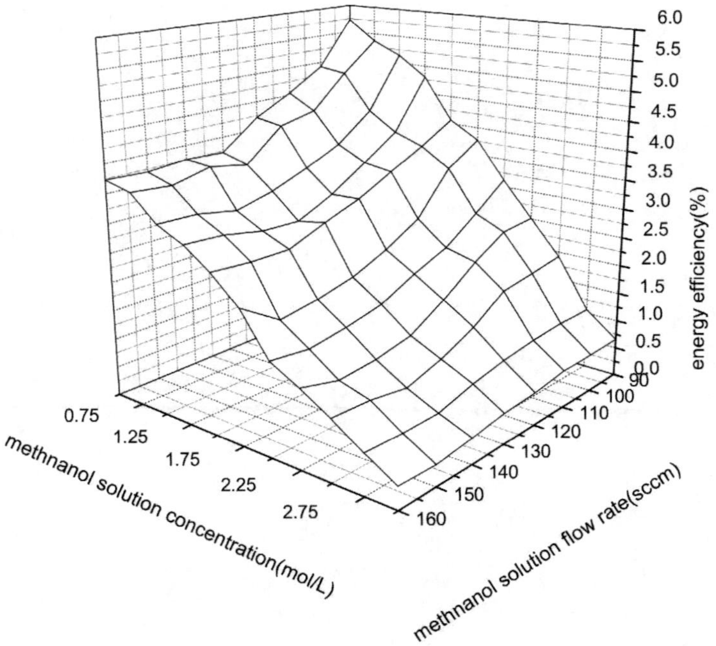
energy efficiency(%)
methnanol solution concentration(mol/L)
methnanol solution flow rate(sccm)
0.0
0.5
1.0
1.5
2.0
2.5
3.0
3.5
4.0
4.5
5.0
5.5
6.0
0.75
1.25
1.75
2.25
2.75
90
100
110
120
130
140
150
160

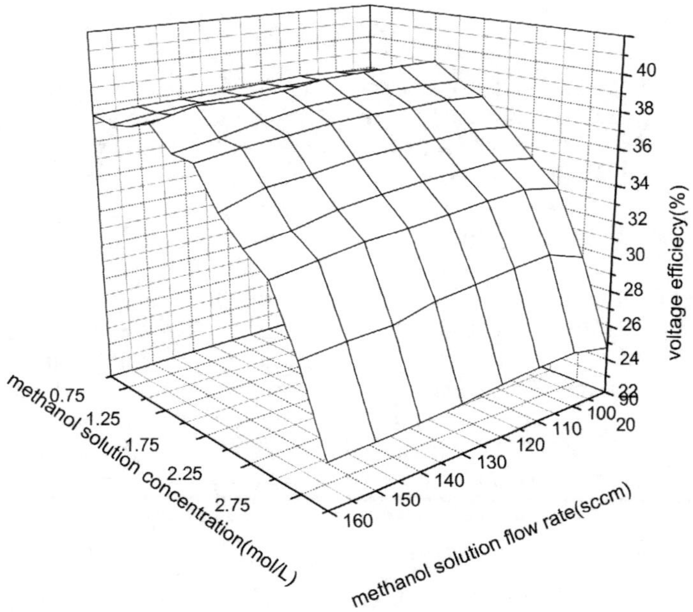
voltage efficiecy(%)
methanol solution concentration(mol/L)
methanol solution flow rate(sccm)
20
22
24
26
28
30
32
34
36
38
40
0.75
1.25
1.75
2.25
2.75
90
100
110
120
130
140
150
160

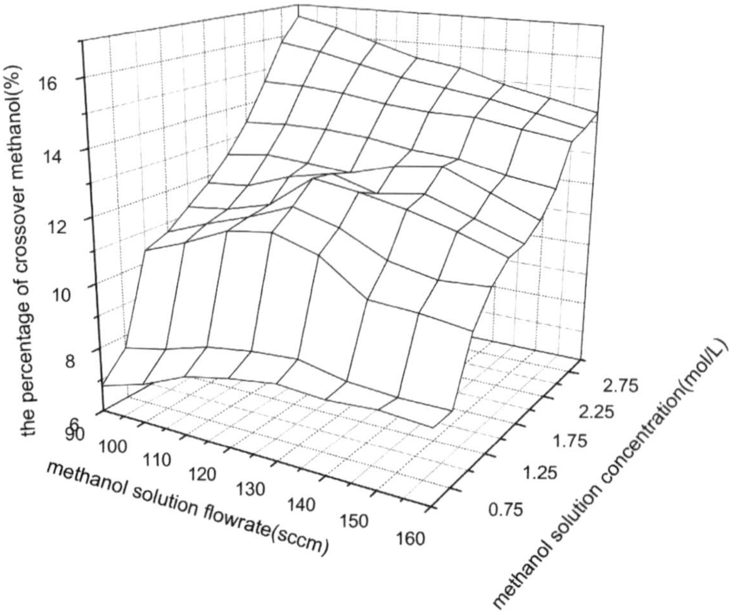

Fig.4 Effect of operation conditions (methanol concentration and methanol flow rate) on the stack efficiency

4 .Conclusions

The distribution of the individual cell's performance in the stack strongly depends on the operating conditions, especially on the supply of the reactants. The higher flow rates of the reactants, methanol and air may ensure a uniform supply of the reactants over all the cells. Therefore, the optimum operating condition is a compromise between the uniform supply of the reactants and the temperature. By optimizing the operating conditions, we could achieve a maximum power of 80W with the 40-cell stack under 120 sccm of 1M methanol and 12 slm of dry air, at room temperature and atmospheric pressure.

A simple engineering method for evaluating the stack's health is provided by the way that determines the maximum power output at different currents. We can also evaluate the *R* or *b*. The stack's efficiency is determined by measuring the CO2 amounts at the anode and cathode outlet. In comparison, equivalent "carbon-flow current" is defined. Results indicate that both the methanol crossover and the interior circuit in the stack contribute to fuel loss. Low fuel flow-rates and low concentrations enable very high fuel efficiencies (i.e. energy densities). However, mass transport limitations prevent the attainment of useful power densities under those conditions that enable high fuel efficiencies.

Acknowledgement

The work is funded by the National High Technology R&D Program of China (2006AA05Z138), and Natural Science Foundation of China (50573041,50703021).

References

1. A. Oedegaard, J. Power Sources, 157 (2006) 244-252
2. V. S. Silva, S. Weisshaar, R. Reissner, B. Ruffmann, S. Vetter, A. Mendes, L.M. Madeira, S. Nunes, J. Power Sources, 145 (2005) 485-494
3. Y. Xu, X. Xie, J. Guo, S. Wang, Y. Wang, V.K. Mathur, J. Power sources, 162(2006) 132-140

4. B. Bae, B. K. Kho, T. H. Lim, I. H. Oh, S.A. Hong, H. Y. Ha, J. Power sources, 158 (2006) 1256-1261
5. C. Xie, J.Bostaph, J. Pavio, J. Power Sources, 136 (2004) 55-65
6. T. Schultz, K. Sundmacher, J. Power Sources, 276 (2006) 272-285
7. R. Dillon, S. Srinivasan, A.S. Aricò, V. Antonucci, J. Power Sources, 127 (2004) 112-126
8. X. Ren, P. Zelenay, S. Thomas, J. Davey, S. Gottesfeld, J. Power Sources, 86 (2000) 111-116
9. C.K. Dyer, J. Power Sources, 106 (2002) 31-34
10. K. Cowey, K.J. Green, G.O. Mepsted, R. Reeve, Current Opinion in Solid State and Materials Science, 8 (2004) 367–371
11. S. Surampidi, S.R. Narayanan, E. Vamos, H. Frank, G. Halpert, A. LaConti, J. Kosek, G. K. Surya Prakash, G. A. Olah, J. Power Sources, 47 (1994) 377-385
12. D.H. Jung, C. H. Lee, C.S. Kim, D.R. Shin, J. Power Sources, 71 (1998) 169-173
13. N. Nakagawa, Y. Xiu, J. Power Sources, 118 (2003) 248-255
14. J. Ge, H. Liu, J. Power sources, 142 (2005) 56-69
15. D. Kim, J. Y. Lee, T. H. Lim, I. H. Oh, H. Y. Ha, J. Power Sources, 155 (2006) 203-212
16. Y. Liu, X. Xie, Y. Shang, R. Li, L. Qi, J. Guo, V. K. Mathur, J. Power Sources, 164 (2007) 322-327
17. D. Buttin, M. Dupont, M. Straumann, M. Straumann, R. Gille, J. C. Dubois, R. Ornelas, G. P. Fleba, E. Ramunni, V. Antonucci, A.S. Arico, P. Creti, E. Modica, J. P. Ganne, J. Appl. Electrochemistry, 31 (2001) 275-279
18. H. Dohle, H. Schmitz, T. Bewer, J. Mergel, D. Stolten, J. Power Sources, 106 (2002) 313-322
19. V. Gogel, T. Frey, Y. Zhu, K. A. Friedrich, L. Jorissen, J. Garche, J. Power Sources, 127 (2004) 172-180
20. D. Chu, R. Jiang, Electrochemica Acta, 51 (2006) 5829-5823
21. R. Jiang, C. Rong, D. Chu, J. Power Sources, 126 (2004) 119-124
22. J.G. Liu, T.S. Zhao, Z.X. Liang, R. Chen, Journal of Power Sources, 153 (2006) 61–67
23. H. Dohle, J. Mergel, D. Stolten, Journal of Power Sources, 111 (2002) 268–282
24. R. Li, MS Thesis, Tsinghua University, Beijing, China, 2007

173g Application of Ferric Sulfate as Catalyst for Conversion of Free Fatty Content of Acid Oil to Biodiesel

Bharat S. Mhatre[1], Dr. Sanjay M. Mahajani[2], Shlok Malik[1], and Varun Mangalvedi[1]. (1) Department of Chemical Engineering, Thadomal Shahani Engineering College, Mumbai University, Mumbai, India, (2) Department of Chemical Engineering, Indian Institute of Technology, Mumbai, India

Biodiesel is an alternative fuel for diesel engines consisting of the alkyl monoesters of fatty acids from vegetable oils. Currently most of the biodiesel is produced from transesterification reaction. This reaction requires vegetable oil, methanol, and an alkaline catalyst. But, the high cost of vegetable oil, its importance as a food product and presence of high content of FFA makes cost-effective production of biodiesel very challenging.

In our research, we are investigating the conversion of the free fatty acids (FFA) content of waste acid-oils from refineries to biodiesel and related kinetics. The problem with processing these acid-oils containing large amounts of FFAs that cannot be converted to biodiesel using an alkaline catalyst. In this paper, we propose application of ferric sulphate as acid catalyst to esterify the free fatty acids before transesterifying the triglycerides with an alkaline catalyst to complete the reaction. Therefore, the synthesis of biodiesel now consists of two-step acid-base catalysis.

The paper discusses the reaction kinetic analysis of the above-mentioned acid catalyzed step to finalize the catalyst loadings, methanol proportions and the reaction time for obtaining optimum conversion.

170a Investigation of Gas-Liquid Slug Formation in a Microchannel T-Junction

Rafael Santos, Department of Chemical Engineering and Applied Chemistry, University of Toronto, 200 College Street, Toronto, ON M5S 3E5, Canada and **Masahiro Kawaji**, Chemical Engineering and Applied Chemistry, University of Toronto, 200 College Street, Toronto, ON M5S 3E5, Canada.

Gas-liquid two-phase flow in a microfluidic T-junction with nearly square microchannels of 113 μ m hydraulic diameter was investigated experimentally and numerically for applications to microreactors and micro heat exchangers. Air and water were used as the working fluids and their superficial velocities were 0.018–0.791 m/s and 0.042–0.757 m/s, respectively. Three-dimensional modeling was performed with computational fluid dynamics (CFD) package FLUENT and the volume-of-fluid (VOF) model. 3D modeling resolved surface tension effects more accurately than 2D simulation. Slug flow (snapping/ breaking/jetting) and stratified flow were observed experimentally. From numerical results void fraction was found to follow a linear relationship with the homogeneous void fraction; two-phase frictional pressure drop agreed with the Lockhart-Martinelli correlation with a Chisholm constant of 5; and the contact angle value affected the shape and size of gas slugs and determined the occurrence of liquid film and corner flow. Correlations were developed for prediction of flow parameters. Higher experimental velocity slip caused by gas inlet pressure build-up can result in deviation from numerical predictions.

170b Miniaturization Aspects of Thermodynamic Cycle Processes for Mobile Energy Generation

Norbert Kockmann, Process Optimization, Lonza, Lonzastr., Visp, 3930, Switzerland

Conventional energy generation is mainly done in thermodynamic circle processes such as steam power process (Brayton or Clausius-Rankine process) or gas turbine process (Joule process). For smaller units, piston machines are used. Since nearly a decade, there are attempts to miniaturize thermodynamic processes for mobile energy supply.

This contribution gives an overview about the recent activities in microstructured thermodynamic cycles. The limits by heat transfer, maintenance of temperature differences and viscous dissipation will be encountered against the gain of these devices.

170c Enhanced Gas Turbine Process with Microstructured Devices

Norbert Kockmann, Process Optimization, Lonza, Lonzastr., Visp, 3930, Switzerland

The gas turbine process (Joule process) is one of the most common thermodynamic cycles for energy generation. The thermodynamic efficiency is determined by the highest process temperature as well as the isentropic efficiencies of compressor and turbine. While major effort was put on material enhancement of turbine blades and aerodynamic design of compressors stages, microstructured devices open new opportunities for process augmentation. This presentation displays application possibilities of micromixers in premixed combustion for compact combustion chambers and the integration of intermediate mixing steps for constant temperature in the turbine. The influence on the overall efficiency is shown. Additionally, some possibilities of oxygen enrichment by microstructured devices are shown to reduce the compression effort.

170d Experimental and Analytical Study of the Motion of a Liquid Slug and Ferrofluid in a Microchannel

Yuichi Shibata[1], Tomoaki Okano[1], and **Masahiro Kawaji**[2]. (1) Mechanical Engineering, Ibaraki National College of Technology, Hitachinaka, 866, Japan, (2) Chemical Engineering and Applied Chemistry, University of Toronto, 200 College Street, Toronto, ON M5S 3E5, Canada

In various microfluidic devices, controlling the flow rate of liquid or gas accurately at micro or nanoliter volume levels is required. By using a ferrofluid and electromagnet, the flow of a liquid or gas in a microchannel can be controlled by the magnetic force exerted on the ferrofluid. In this study, an unsteady motion of a liquid slug caused by the driving force from the ferrofluid has been investigated in a 200μ m circular microchannel. The motion and behavior of the liquid slugs were observed. The surface tension and kinematic viscosity of the liquid were found to significantly affect the movement of the liquid slug. The velocity of the ferrofluid was in turn affected by the physical properties of the liquid pulled. Combinations of various liquids with the ferrofluid were examined in this study. The velocity of the ferrofluid was found to vary with the dynamic contact angle of the liquid slug. The movement and behavior of the interface between two different liquids were investigated experimentally and analytically. The interfacial shape changed continuously until the bidrops started to move. In addition, the interfacial tension of bidrops was estimated using the interfacial contact angle. We could relate the characteristics of the movement of bidrops to the interfacial variation.

199a Toxic Response of Clostridium Beijerinckii and Escherichia Coli to Lignin Hydrolysate-Related Compounds

Robert J. Mitchell[1], Yun-Jie Lee[1], Sun-Mi Lee[1], Joo-Myung Ahn[2], Dae Haeng Cho[3], Young-soon Um[1], Yong Hwan Kim[3], Man Bock Gu[4], and Byoung-In Sang[1]. (1) Center for Environmental Technology Research, Korea Institute of Science and Technology (KIST), Seoul, South Korea, (2) Dept of Biotechnology, Korea University, Seoul, South Korea, (3) Dept. of Chemical Engineering, Kwangwoon University, Seoul, 139-701, South Korea, (4) Department of Biotechnology, Korea University, Seoul, South Korea

The toxicity of several compounds commonly found in lignin hydrolysates were evaluated individually by monitoring the production of acetone, butanol and ethanol (ABE) from Clostridium beijerinckii. The compounds selected for analysis included coumaric acid, syringaldehyde, furfural, ferulic acid and glucuronic acid. Assays with C. beijerinckii were performed using batch cultures while the growth and production of each of the ABE compounds were assessed over a 72-hour period. Very significant losses in butanol and acetone production were evident when the cultures were exposed to 0.5 mg/ml coumaric or ferulic acid, with a concurrent increase in the amount of acetic acid and butyric acid being produced. Likewise, exposure to syringaldehyde led a reduced butanol production while the addition of up to 1mg/ml of furfural or glucuronic acid had little or no effect on the metabolic output of this strain. To further evaluate, and potentially identify, the stresses C. beijerinckii experience when exposed to each of these compounds, tests were performed using a panel of bioluminescent strains of E. coli, each of which carry a plasmid harboring a specific promoter fusion with the luxCDABE genes. The strains included in this study are specific for and respond to oxidative damage (strains DK1 (katG) and DS1 (sodB)), membrane damage (strain DPD2540 (fabA)), protein damage (TV1061 (grpE) and DC1 (clpB)) and DNA damage (DPD2794 (recA)). Two other strains that were included were GC2 (lacZ) and DNT5 (nagR-nagAa), which are used to evaluate for the overall cytotoxicity and the presence of salicylic acid derivatives, respectively. Therefore, the stresses experienced when bacteria are exposed to each compound were characterized based upon the bioluminescent responses from these strains. Syringaldehyde clearly leads to a significant amount of membrane damage, with the bioluminescent response from strain DPD2540 being induced more than 400-fold when syringaldehyde was added to a concentration of 1mg/ml. Interestingly, glucuronic acid had little effect on C. beijerinckii but led to a significant response (about 200-fold with a concentration of 1.56 mg/ml) from DPD2540, suggesting that it does lead to some membrane damage. Furthermore, a slight but significant response (2.6-fold) was seen from strain DK1 when glucuronic acid was added at a concentration of 0.78 mg/ml, while higher concentrations led to a reduced response.

Control of Ethanol Dehydration in the Supercritical Water Reforming of Ethanol into Hydrogen

Jonathan E. Wenzel, Jason W. Picou and Sunggyu Lee, Missouri University of Science and Technology

Abstract

Ethanol, a renewable resource, may be reformed non-catalytically into hydrogen by a novel process utilizing supercritical water which acts synergistically as a solvent and as a reactant. Ethanol reformed by this process does not require energy-intensive separation following fermentation or subsequent distillation. In addition this novel process produces little or no carbon build-up inside the reactor even after days of continuous operation. The principal reactions that occur are: (1) direct reformation of ethanol into hydrogen and carbon oxides, (2) pyrolytic decomposition of ethanol into hydrogen, methane, and carbon oxides, and (3) dehydration of ethanol. Dehydration of ethanol occurs as an undesired reaction in the non-catalytic reformation of ethanol to hydrogen. A preliminary series of supercritical water reformation of ethanol experiments were conducted to validate a novel 400 mL Haynes Alloy 230 tubular reactor for supercritical water reformation process. It is found that the ethanol dehydration reaction is present as a side reaction even in the water-rich mixture of the current process and its byproduct ethylene further undergoes hydrogenation reaction with the process ultimate product, hydrogen, producing ethane. Experiments were performed between 857 K and 1042 K at 24 MPa. The effects of temperature and space time upon molar concentration of ethylene and ethane in the gaseous ambient products are discussed. The optimal process conditions that minimize this wasteful ethane formation are found and experimentally demonstrated.

Introduction

Ethanol is a widely available chemical feedstock produced from domestic sources that is a gateway fuel towards energy independence in the United States. Ethanol is commonly used as an oxygenated additive in gasoline in replacement of MTBE for air quality control. An alternative means of utilizing ethanol as an energy source is converting ethanol into hydrogen for use in internal combustion engines and fuel cells. Ethanol is conventionally reformed into hydrogen by means of steam-reforming in the presence of a catalyst. Catalytic reformation of ethanol is susceptible to catalyst poisoning by feedstock impurities, catalyst degradation, coking, and the production of several undesirable products in addition to hydrogen.[1] A novel, non-catalytic, approach to reforming ethanol into hydrogen is to use supercritical water as a reaction medium. In this approach supercritical water acts synergistically both as a solvent and as a reactant.[2] This process is not susceptible to many of the pitfalls of catalytic reformation and was originally developed for the reformation of military logistic fuel (JP-8) to hydrogen.[3] Another advantage supercritical water reformation has over catalytic steam reforming is that supercritical water is denser by orders of magnitude than steam requiring smaller reactor volumes than catalytic reactors.

The reformation products of supercritical water reformation of ethanol include hydrogen, carbon oxides, methane, and ethane. This process completely converts ethanol into gaseous products with no noticeable coking or fouling and with only trace amounts of carbonaceous species in the liquid effluent. In addition most of the carbon oxides are in the form of carbon dioxide thus eliminating the need for a separate water gas shift reactor. In previous work, using a 1-liter tubular Inconel 625 Grade 1 reactor, three principle reactions were determined to occur in the supercritical water reformation of ethanol into hydrogen: (1) direct reformation of ethanol into hydrogen and carbon oxides, (2) pyrolytic decomposition of ethanol into hydrogen, methane, and carbon oxides, and (3) dehydration of ethanol.[2] Kinetic parameters were also regressed for the direct reformation reaction and the pyrolytic decomposition reaction.[4] The third, undesired reaction, dehydration of ethanol to form ethylene and subsequent hydrogenation of ethylene to form ethane was studied in further detail as part of the validation of a novel 400 mL Haynes Alloy 230 tubular reactor for supercritical water reformation of military logistic fuel.

Background

Water is in a supercritical state when its pressure is greater than 22.064 MPa and its temperature is greater than 374°C, as shown in Figure 1.[5] Above these conditions water exhibits both liquid and gas-like properties with heat capacity, thermal conductivity, viscosity, and electrolytic conductivity varying with temperature and pressure. In addition supercritical water has a low dielectric constant in comparison to ambient water, making it a non-polar solvent. As a result water can form a Type I supercritical mixture with ethanol at temperatures and pressures below water's critical point and a continuous critical mixture curve, Figure 2.[6]

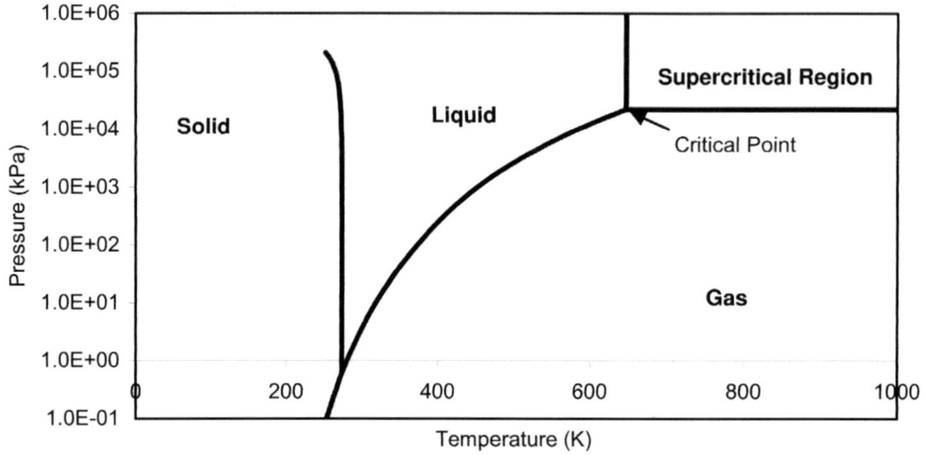

Figure 1: *Phase diagram for water.*

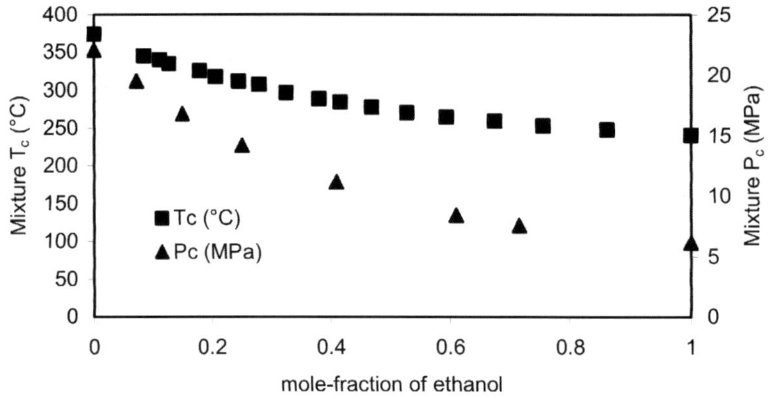

Figure 2: *Experimentally determined critical points of ethanol-water mixtures.*

There are three principal concurrent chemical reactions that occur in supercritical water reformation of ethanol: direct supercritical water reformation, pyrolytic decomposition of ethanol, and dehydration of ethanol. The direct supercritical water reformation reaction is

$$C_2H_5OH + 3\ H_2O \rightarrow 2CO_2 + 6H_2 \quad (1)$$

On the other hand, the decomposition of ethanol is the summation of the dehydrogenation of ethanol (2) to form acetaldehyde and the decarbonylation of acetaldehyde (3) to form methane and carbon monoxide. Carbon monoxide is then subsequently converted into carbon dioxide by the water gas shift reaction:

dehydrogenation: $C_2H_5OH \rightarrow C_2H_4O + H_2$ (2)
decarbonylation: $C_2H_4O \rightarrow CH_4 + CO$ (3)
water gas shift: $CO + H_2O \rightarrow CO_2 + H_2$ (4)

This reduces to an overall stoichiometric equation:

$$C_2H_5OH + H_2O \rightarrow CH_4 + CO_2 + 2H_2 \quad (5)$$

In addition, ethanol can be dehydrated to form ethylene which subsequently reacts with hydrogen to form ethane:

dehydration: $C_2H_5OH \rightarrow C_2H_4 + H_2O$ (6)
hydrogenation: $C_2H_4 + H_2 \rightarrow C_2H_6$ (7)

According to Arita et al. (2003)[7], at 773 K in a sealed quartz reactor the principal products of supercritical water reformation of ethanol are acetaldehyde, methane, carbon oxides, ethylene, and ethane. However, the decarbonylation reaction occurs very quickly, in supercritical water reformation in an Inconel 625 Grade 1 reactor acetaldehyde concentrations become negligible[2] above 873 K and in catalyzed systems acetaldehyde concentrations become negligible above 923 K.[8]

Experimental Section

Chemicals

The feed ethanol used for this work was 95% ethanol by volume with the balance being water. The feed water used for this work was distilled and deionized. The carrier gas for the gas chromatograph was Ultra Pure Carrier Grade (UPC 5.5) Argon with a purity of 99.9995%.

Apparatus

Experiments were conducted using a custom-designed Haynes® Alloy 230 supercritical water reformation reactor system. This system consists of a feed system, integrated heat exchanger for heat recovery, reactor, sample collection system, and data acquisition and control system. A schematic process flow diagram is shown in Figure 3. The feed system consists of micrometering pumps for water and ethanol and provides accurate feed-rate measurements for material balances. The reactor is tubular with an internal volume of 383 mL and constructed of Haynes® Alloy 230 permitting supercritical water experiments to be conducted at pressures and temperatures up to 36.3 MPa at 800°C. Various sampling systems allow for the collection of gaseous and liquid effluent from the reactor at ambient conditions for compositional analysis and material balances. A wet-test meter is used for determining the gaseous effluent flow rate.

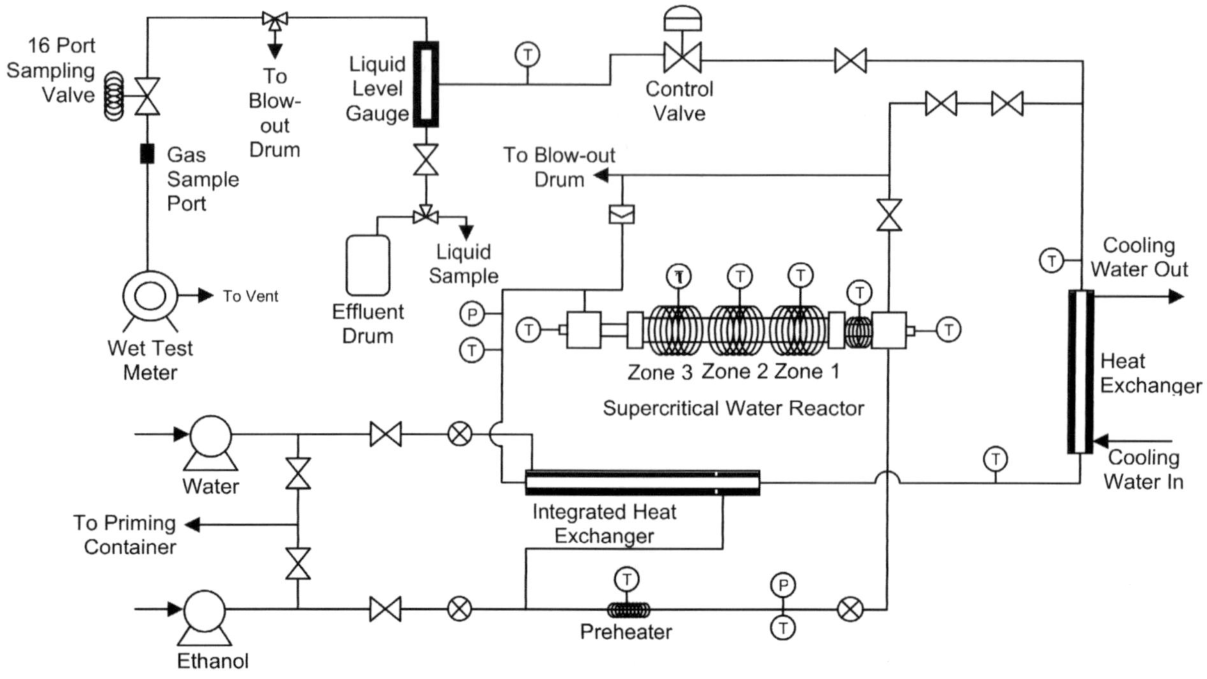

Figure 3: *A schematic of supercritical water reformation system at MS&T.*

Analysis of the gaseous effluent was performed using a HP 5890 Series A gas chromatograph with a thermal conductivity detector. Liquid effluent was analyzed for total carbon content using a Dohrmann DC-190 TOC.

Procedure

Experiments were performed according to the following procedure: start-up of the system to operating conditions, sample and data collection, and shut-down of the system. To start-up the system, the supercritical water reactor (SWR) was heated and once the internal temperature of SWR approached the operating temperature, water was fed into the reactor. Once the SWR inlet temperature and integrated heat exchanger temperature stabilized, the experimental run started by feeding ethanol into the reactor. Depending on the experimental run's residence time, the production of gaseous effluent would reach a steady flow-rate half an hour after ethanol was fed into the reactor. During an experiment, multiple gas and liquid samples were collected and reactant fed rates and effluent flow rates were recorded. The data acquisition and control software recorded temperature and pressure at various locations in the system. At the conclusion of an experimental run, additional experiments were conducted by changing experimental conditions and allowing the system to stabilize. To shut-down the system, the reactor was slowly depressurized and allowed to cool.

Results and Discussion

As part of an evaluation of the Haynes® Alloy 230 reactor, seven ethanol reformation experiments were performed. The experiments were carried out between 850 and 1050 K at a constant pressure of 24.2 MPa. At the conclusion of the experiments the reactor was opened for inspection and found to be free of carbon deposits. The test conditions, ambient reactor effluent gas flow rate, and compositions on a water free basis are listed in Table 1. Table 2 lists the mole fractions of the reactor effluent.

Table 1: Supercritical Water Reformation Conditions and Ambient Gas Effluent Concentrations

Run ID	T / K	EtOH / $g\cdot min^{-1}$	H_2O / $g\cdot min^{-1}$	P / MPa	Gas Rate / $L\cdot min^{-1}$	H_2 / mol-%	CO / mol-%	CH_4 / mol-%	CO_2 / mol-%	C_2H_6 / mol-%	C_2H_4 / mol-%
WE_135	858	0.7	18.8	24.0	0.35	30.8%	26.3%	27.5%	1.7%	9.3%	4.2%
WE_137	896	0.9	20.5	24.2	0.90	25.4%	34.7%	31.2%	1.1%	11.2%	0.7%
WE_138	911	0.9	18.9	24.2	1.10	26.2%	32.9%	29.4%	2.9%	11.1%	0.0%
WE_139	960	0.9	20.1	24.2	1.22	34.4%	20.1%	30.1%	10.5%	9.1%	0.0%
WE_140	1020	0.8	20.8	24.2	1.49	41.6%	4.7%	29.6%	18.8%	4.6%	0.0%
WE_143	1042	0.9	18.8	24.1	1.48	41.9%	1.4%	30.5%	21.1%	2.3%	0.0%
WE_141	1019	0.4	10.5	24.2	0.71	42.4%	1.2%	30.1%	21.5%	3.1%	0.0%

Table 2: Supercritical Water Reformation Feed and Effluent Mole Fractions Acetaldehyde Free Basis

	Reactor Feed			Reactor Effluent Acetaldehyde Free Basis						
Run ID	T / K	EtOH / mol-%	H_2O / mol-%	H_2O / mol-%	H_2 / mol-%	CO / mol-%	CH_4 / mol-%	CO_2 / mol-%	C_2H_6 / mol-%	C_2H_4 / mol-%
WE_135	858	1.37%	98.63%	98.69%	0.40%	0.34%	0.36%	0.02%	0.12%	0.06%
WE_137	896	1.61%	98.39%	96.86%	0.77%	1.04%	0.94%	0.03%	0.34%	0.02%
WE_138	911	1.77%	98.23%	95.88%	1.05%	1.32%	1.18%	0.12%	0.45%	0.00%
WE_139	960	1.76%	98.24%	95.69%	1.42%	0.83%	1.25%	0.43%	0.38%	0.00%
WE_140	1020	1.49%	98.51%	95.12%	2.04%	0.23%	1.46%	0.92%	0.23%	0.00%
WE_143	1042	1.44%	98.56%	95.42%	1.98%	0.06%	1.40%	1.00%	0.14%	0.00%
WE_141	1019	1.76%	98.24%	94.76%	2.26%	0.08%	1.64%	1.14%	0.12%	0.00%

It is important to note that one of the test conditions, Run ID WE_135, is below 873 K. In previous supercritical water reformation studies at temperatures between 773 K and 873 K, ethanol was completely dehydrogenated to form acetaldehyde though there remained detectable quantities of acetaldehyde since the decarbonylation of acetaldehyde did not proceed to completion.[2, 7] Based on an atomic carbon balance in Table 3 between the carbon fed as ethanol and the carbon exiting the reactor as gaseous species at ambient conditions, and the lack of a solid product, it is evident that a there is a liquid product collected ambient conditions for Run WE_135, acetaldehyde. In addition, total carbon analysis of the liquid effluent of Run ID WE_137 had a total carbon content of 2274 ppm, which is indicative that acetaldehyde was not completely converted by the decarbonylation reaction. Closure of the carbon balance at higher temperatures suggests that the decarbonylation reaction has proceeded to close to completion.

The presence of ethylene and ethane are indicative that ethanol is being dehydrogenated, which is remarkable given that the molar ratio of water to ethanol is approximately 60 for the experiments considered. The production of ethylene and ethane as a by-product in the supercritical water reformation of ethanol is a concern due to multiple reasons. First, ethanol is being consumed in reactions that do not produce hydrogen. Second, ethylene could potentially be polymerized producing coke and fouling a reactor. In addition when ethylene is hydrogenated to ethane, hydrogen, the desired product, is consumed. Hence, it is desirable to limit the amount of ethylene and ethane produced. It should be also noted that both ethylene and ethane can be reformed to generate hydrogen in the supercritical water reformation process; however, its kinetic rate is slower than that of the ethanol reformation at the conditions studied in this paper.

Table 3: Atomic Carbon Balance on Carbon exiting the Reactor as a Gaseous Species at Ambient Conditions

Run ID	T / K	Carbon In mmol·min^{-1}	Carbon Out mmol·min^{-1}	Closure
WE_135	858	29.0	11.5	40%
WE_137	896	37.4	32.4	87%
WE_138	911	37.7	38.5	102%
WE_139	960	40.1	38.2	95%
WE_140	1020	35.0	36.9	105%
WE_143	1042	17.0	16.7	98%
WE_141	1019	37.4	34.0	91%

At temperatures of 896 K and below, detectable amounts of ethylene are observed, whereas at higher temperatures ethylene concentrations are below detection limits. When comparing runs WE_135 and WE_137 it is apparent that the hydrogenation of ethylene to ethane is less active at lower temperatures. It may be possible that either the hydrogenation reaction is less active due to lower temperatures or at the lower temperature experiments the mole fraction of hydrogen was lower (Table 2). It is likely that both circumstances are contributing to the detectable presence of ethylene which merits further study. In experiments at temperatures of 911 K and above, all the ethanol fed is converted to gaseous products and ethylene is completely hydrogenated into ethane. In addition at higher temperatures more gaseous effluent is produced as shown in Table 1 and ethane is less abundant. This is indicative that ethanol is preferentially being reformed instead of being dehydrated. The effect of temperature upon the combined mole fractions of ethylene and ethane are shown in Figure 4.

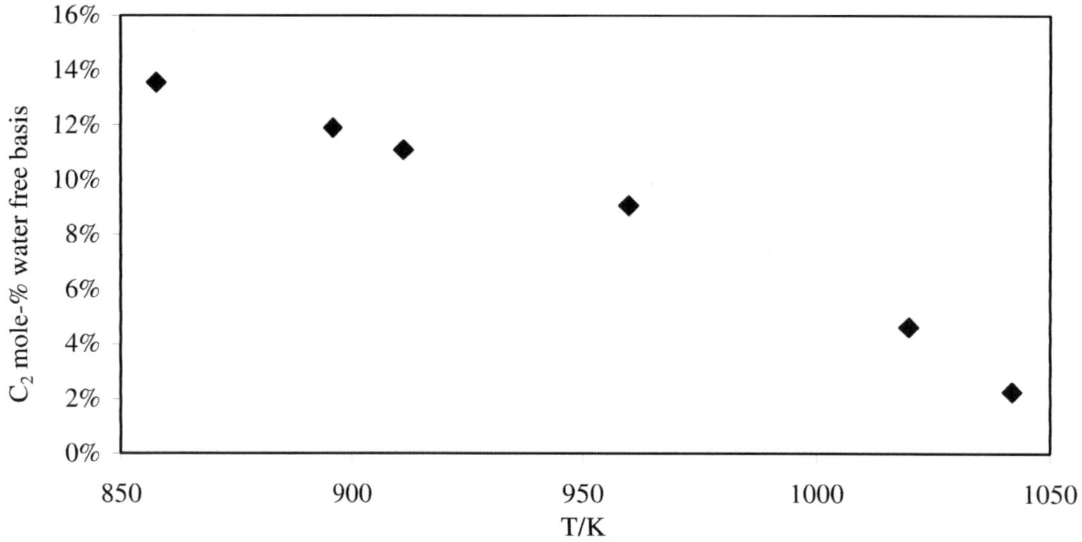

Figure 4: *Combined mole fraction of ethylene and ethane versus temperature.*

Experimental Runs WE_140 and WE_141 provide a comparison of the effect of residence time upon conversion of ethanol in supercritical water reformation. Run WE_140 has approximately half the residence time of WE_141. At an increased residence time hydrogen content of the reactor effluent increases while ethane content of the reactor effluent decreases. This may be attributable to the contributions of the ethane reformation that takes place simultaneously in the reactor, thereby deceasing the ethane concentration while increasing the hydrogen yield. Kinetic studies to determine the effect of residence time and temperature in a Haynes® Alloy 230 reactor would provide considerable insight on minimizing the ethanol dehydration reaction while maximizing ethanol reformation.

Conclusions

Supercritical water is used as a solvent and as a reactant in the non-catalytic reformation of ethanol into hydrogen. Three principal reactions that occur in the supercritical water reformation of ethanol are (1) direct reformation into hydrogen and carbon dioxide, (2) pyrolytic decomposition of ethanol into methane, carbon dioxide, and hydrogen, and (3) dehydration of ethanol. The supercritical water reformation of ethanol was studied in order to produce hydrogen from renewable feedstocks such as corn ethanol and also as part of the validation trials of a novel supercritical water reactor constructed of Haynes® Alloy 230. It was found that the undesired competing reaction in supercritical water reformation, the dehydration of ethanol is less pronounced as temperature is increased in relation to the reformation reactions when the reactant feed rates and reactor pressure are kept constant. Also an increase in reactor residence time may decrease the selectivity of the dehydration of ethanol over that of the reformation of ethanol to hydrogen.

Acknowledgments

This work was supported by a grant from National University Transportation Center (NUTC), U.S. Department of Transportation. The major parts of the experimental hardware system were developed using financial support from DRS Technical Services, Inc. The authors are also grateful for helpful comments provided by Dr. H. Bryan Lanterman.

References

1. Morgenstern, D. A.; Fornango, J. P., Low-temperature reforming of ethanol over copper-plated raney nickel: a new route to sustainable hydrogen for transportation. *Energy & Fuels* 2005, 19, 1708-1716.

2. Wenzel, J.; Picou, J.; Factor, M.; Lee, S., Kinetics of Supercritical Water Reformation of Ethanol to Hydrogen. In *Materials Science & Technology 2007 Conference and Exhibition*, Detroit, Michigan, 2007.

3. Lee, S.; Lanterman, H. B.; Wenzel, J. E.; Edwards, N.; Adams, A.; Wootton, J.; Garcia, A., Noncatalytic direct reformation of JP-8 fuel in supercritical water. *Preprints - American Chemical Society, Division of Petroleum Chemistry* 2006, 51, 487-490.

4. Wenzel, J. E.; Niemoeller, A.; Stever, M. S.; Lee, S., Kinetics of Reforming Ethanol into Hydrogen in a Supercritical Water Medium. In *2007 AIChE Annual Meeting*, Salt Lake City, Utah, 2007.

5. IAPWS *Release on the pressure along the melting and the sublimation curves of ordinary water substance*; The International Association for the Properties of Water and Steam: Milian, Italy, September, 1993; pp 1-4.

6. Hicks, C. P.; Young, C. L., The gas-liquid critical properties of binary mixtures. *Chem. Rev.* 1975, 79, 119-175.

7. Arita, T.; Nakahara, K.; Nagami, K.; Kajimotoa, O., Hydrogen generation from ethanol in supercritical water without catalyst. *Tetrahedron Letters* 2003, 44, 1083-1086.

8. Deluga, G. A.; Salge, J. R.; Schmidt, L. D.; Verykios, X. E., Renewable Hydrogen from Ethanol by Autothermal Reforming. Science 2004, 303, (5660), 993-997.

Dilute Acid Prehydrolysis and Extraction of Hemicellulose Prior to Aspen Chemi-Thermomechanical Pulping

Yaman Boluk, Zhirun Yuan, Frank Tosto, **Thomas Browne**, Bruce Atkinson

1. Introduction

Hemicellulose, accounting for an average 22 % of hardwood, is usually underutilized or somewhat wasted in various pulping processes. For example, approximately 2-5% of hemicelluloses are dissolved and dispersed into process water in a mechanical pulping process. Such released hemicelluloses are accumulated in the process water and often interfere with papermaking processes. In addition they also increase the effluent load and contribute to the treatment cost.

Hemicelluloses can be extracted from wood prior to mechanical pulping. This process not only offers new feedstock for the production of chemicals and fuels in the mill but also remedies some of the operational problems in pulp and paper mills. However it is important to keep in mind that certain physical properties of pulp require hemicellulose in the fiber matrix. The desired amount of hemicellulose needed in the pulp product depends entirely on the wood species, pulping process and end use of the pulp. Therefore while developing the hemicellulose pre-extraction process, one should consider not only the yield and composition of extracted hemicellulose but also the pulp properties produced from hemicellulose pre-extracted wood.

Much of the work on the hydrolysis of hardwoods and the solubilisation of hemicelluloses have been done by steam and dilute acids treatments (1-3). Although a wealth of information exists on the removal of hemicelluloses and investigating the kinetics of extraction and composition of extracted sugars, there has not been any pilot plant study to investigate mechanical pulping and pulp properties in combination with hemicellulose pre-extraction process.

In this study aspen (*Populus tremuloides*), which is a widely utilized hardwood in the bleached chemi- thermomechanial pulping (BCTMP) process has been chosen as a feedstock. Our goal is to extract hemicelluloses from wood while still producing pulp with excellent physical properties. This undoubtedly requires not only high-efficiency hemicellulose extraction process but also the optimization of hemicellulose pre-extraction process balancing the removal of hemicellulose and retention of physical properties of pulp. In this preliminary study we have investigated in a limited fashion the pre-hydrolysis of aspen prior to BCTMP process and pulp properties after the mechanical pulping and bleaching of hemicellulose-extracted wood. In addition to the pilot scale study, BCTMP mill pressate samples were also collected for further chemical composition analysis and purification.

2. Materials and Methods

2.1. Sample Analysis

Aspen wood chips were prepared and moisture content was determined according to TAPPI T264 cm-97 before the chemical and carbohydrate analysis. Prepared wood chips were analyzed for ash, lipophilic (ethanol, benzene extractives), water soluble (hot water extractives), lignin, holocellulose, pentosan and carbohydrate contents. Ash amount was determined by following TAPPI method T 413 om-93. Organic soluble extractives were extracted by using an ethanol/benzene extraction procedure (TAPPI T 204 cm-97). Following this, water soluble carbohydrates were determined by following TAPPI T 207 cm-97. The acid-insoluble Klason lignin was determined after ethanol-benzene extraction and a wash with ethanol and hot water by hydrolyzing and solubilising the carbohydrates and filtering the acid insoluble lignin (TAPPI T222 om-98). Pentosans and holocellulose components were analyzed by following TAPPI methods T 223 cm-84 and T 9 w-75 respectively.

The carbohydrate composition of aspen wood chips and other samples were analyzed after acid hydrolysis and reported as monosaccharides. Sulfuric acid is added to make up a 3% H_2SO_4 solution and heated to 116^{0}C for two hours. The sample is cooled and neutralized to approximately pH 4.6. Dionex BIOLC with pulsed amperometeric detector, auto sampler and a data handling system was used for neutral carbohydrate analysis.

2.2. Pulp Mill Pressate

In the mill operation, aspen wood chips are first screened and steamed. The steamed chips are then washed by 60°C water in a chip washer and drained before a three-stage impregnation process: 1st stage with water, 2nd stage with DTPA, and the 3rd stage with 0.8% bisulfite and 0.4% caustic. Pressate samples were collected at chip drainer, before the 1st impregnator (621-138), and before the 3rd impregnator (621-158) for further analysis and to evaluate the recovery and purification potential of dissolved hemicelluloses.

2.3. Pilot Plant Study

2.3.1. Acid Pre-extraction

An Andritz-Bauer 6" MSD press impregnator was used in the pilot study of acid pre-extraction. Aspen chips were impregnated by 0.05 mol/L sulphuric acid at 60ºC. The impregnated chips were cooked at 130ºC for 10 minutes. The cooked chips were fed into the press impregnator again. 20L of pressate was collected.

2.3.2. Pulping and Bleaching

Aspen chips were impregnated by 0.8% bisulfite and 0.4% caustic in the 6" MSD press impregnator. The impregnated chips were retained at 70º for 30 minutes and preheated at 110ºC for 3 minutes before refining. The chips were then refined in two stages with a 22" Andritz pressurized TMP refiner followed by a Bauer 400 36" atmospheric refiner.

Two pulps, one control and one chemical-extracted that had freeness around 350 mL were selected, chelated, and bleached with alkaline peroxide. The bleaching conditions are: 5% NaOH, 6.5-9% peroxide, 1.75% silicate, 0.4% $MgSO_4$, 22% consistency, 70^{0}C, 3hrs. The bleached pulps were tested again.

Refined pulps were hot disintegrated and screened with a 4 cut plate. Pulp properties were tested according to PAPTAC standard methods.

3. Results and Discussion

3.1 Aspen wood composition

The composition of the initial aspen chips sample (ACS) is summarized along with data from the literature (4, 5) in Table 1. ACS holocellulose, lignin and hemicelluloses values are reported on a dry weight and extractive free basis. The variations in reported data are caused in part by variation in the wood composition but also from different analytical methods. It is more likely that holocellulose results reported in the literature included some residual lignin. In fact, the sum of alpha cellulose and polysoses value of 70.6 of Fengel et.al (4) compiled data is comparable ACS holocellulose value.

Table 1. Wood composition of aspen chips

	Holo-cellulose (%)	Alpha Celulose (%)	Poly-soses (%)	Pento-sans (%)	Klason Lignin (%)	Ethanol-benzene extractives (%)	Hot water extractives (%)	Ash (%)
ACS	69.0	n/a	n/a	18.6	17.0	3.0	3.2	1.2
Populus tremuloides(4)	80.3	49.4	21.2	17.2	18.1	3.8	2.8	0.4
Populus tremuloides(5)	78	49	n/a	19	19	3	3	0.4

The predominant polyoses (hemicelluloses) in aspen hardwood are pentosans. The ACS sample contains 18.6% pentosan. Aspen pentosan is a O-acetyl-4-O-methylglucurono-β-(1→4)-D-xylopyranose in short called glucuronoxylan. The backbone consists of β-D-xylopyranose (Xyl) units linked by (1→4) bonds. Most of the xylose units contain acetyl group at C-2 or C-3 with the concentration of seven acetyl residues per ten xylose units. The xylose units in the xylan chain also link by (1→2) bonds to 4-O-methyl-α-D-glucopyranosyl uronic acids (MeGlcA). On average there is one MeGlcA unit per ten Xyl units (6).

In addition to pentosans, aspen harwood contains 2-5% hexosans in the form of glucomannan, with lesser quantities of galactans and glucans (6). Glucomannan is composed of β-D-glucopyranose and β-D-mannopyranose units linked by (1→4) bonds.

The neutral carbohydrate composition of the initial ACS is shown in Table 2, along with another analysis results from the literature (6-8). As discussed in the preceding paragraph, aspen hardwood has seven O-acetyl (MW=43.05) groups and one 4-O-methylglucuronic acid (MW=193.13) group on every ten xylose (MW=150.13) units. Since acetyl and methylglucuronic acid components are not included in the chemical analysis, 34% ((7x43.05+1x208.17)/(10x150.13)) has to be added on xylose weight percent to calculate the total weight of glucuronoxylan. By using this assumption, O-acetyl-4-O-methylglucurono-β-(1→4)-D-xylan accounts for 24% weight of the ACS sample.

Table 2. Neutral carbohydrate composition of o.d. aspen wood

	Glucose (%)	Arabinose (%)	Xylose (%)	Galactose (%)	Mannose (%)	Rhamnose (%)	Total Neutral Sugar (%)
ACS	46.4	0.5	17.9	0.5	2.4	0.5	72.1
Populus tremuloides(4)	43.0	n/a	18.0	1.1	3.0	n/a	65.1
Populus tremuloides(5)	43.5	0.5	16.4	0.6	1.7	0.3	63.0

The glucose:mannose ratio in aspen glucomannan has been reported 1:2 in the literature (6,7). Therefore we can deduct that glucomannan content is approximately 3.5%. The klason lignin content, lipophilic extractives and ash are reported as 17%, 3.0% and 1.2% respectively. The rest of the material can be assumed acid soluble lignin, pectin, and extractives. We can reconstruct the ACS composition as in Table 3. Both glucuronoxylan and glucomannan contents of the aspen chips agree to the amounts usually found in hardwood (6, 9).

Table 3. Estimated composition of aspen chips before treatment

	Cellulose (%)	Gluco mannan (%)	O-acetyl – (methyl glucurono) xylan (%)	Klason Lignin (%)	Acid Soluble Lignin, Pectin and Extract. (%)	Ethanol-benzene extractives (%)	Ash (%)
ACS	45.2	3.6	24	17	5.1	3.0	1.2

Based on reconstructed chemical composition of aspen wood chips, it can be concluded that the theoretical maximum yield of glucuronoxylan residues extraction per 100 g of wood would be 24 g and glucomannan residues would be 3.6 g.

3.2 Removal of Carbohydrates

Hydrolyzed sugar compositions and xylose/glucose ratios of pulp mill and pilot plant pressates are tabulated in Table 4. Although the yield is comparable, acid pretreatment process produces sugars in more concentrated form as opposed to steam impregnation. It can be seen that pilot plant control steam impregnated pressate sample composition and xylose/glucose ratio are in the order of pulp mill pressate samples particularly very close to 622-158 pressate sample. On the other hand acid prehydrolysis resulted in a release of a higher level of xylans than glucans from aspen wood chips. The xylosidic bonds between xylose units are more rapidly hydrolyzed by acid prehydrolysis process than glucosidic bonds in cellulose. Consequently xylose, total sugar concentrations and xylose/glucose ratio were higher in the acid pre-treated pressate.

Table 4. Sugar Analysis of Hydrolyzed Pressate Samples

	Chip Drain Filtrate (CDF)	621-138 Impregnator Pressate	621-158 Impregnator Pressate	Control Steam Impregnation Pressate	Acid Pre-treated Pressate
Glucose, mg/kg	254	515	988	997	477
Xylose, mg/kg	264	184	505	656	3114
Xylose/Glucose	1.04	0.36	0.51	0.66	6.52

3.4 Removal of Acetic Acid

Xylan backbone associated acetyl groups are also hydrolysed and acetyl groups can be released in acetic acid form. As discussed previously aspen hardwood has seven acetyl and one methylglucuronic acid groups on every ten xylose units which can be removed during the prehydrolysis.

Table 5 shows the acetic acid concentration of pressate samples. The acid-prehydrolyzed pressate contains 1010 mg/l acetic acid and 3114 mg/l xylose. On the other hand, control steam impregnated pressate has 167 mg/l acetic acid and 656 mg/l xylose. Theoretically the potential yield of acetic acid is about 6g per 100g wood and xylose yield is about 24g per 100g wood. Therefore the theoretical acetic acid/xylose ratio is 0.25. Both control steam impregnation and acid prehydrolysis processes resulted released acetic acid/xylose ratios around 25-30%.

Table 5. Analysis of Carboxylic Acids in Pressate Samples

Sample	622-138 Impregnator Pressate	622-158 Impregnator Pressate	Chip Drain Filtrate (CDF)	Control Steam Impregnation Pressate	Acid pre-treated pressate
Formic acid, mg/l	258	154	302	47	30
Acetic acid, mg/l	2,820	1,380	3,600	167	1,010

Unlike stem impregnated and acid prehydrolyzed pilot plant pressate samples, pulp mill pressate samples have excessively rich acetic acid concentrations which can be attributed to the mill water recirculation and utilization of bleach plant streams. The presence of formic acid supports this argument.

Glucuronic acid to xylose linkages are generally more resistant to acid hydrolysis than the xylan backbone. Xylan is bound to lignin through uronic acid side chain groups (10). It is assumed that such binding might decrease the removal of uronic acid groups. Therefore it is expected that the removal of uranic groups would be much less than the removal of anhydroxylose units.

3.3 Pulping and Bleaching

Figure 1 shows that acid pre-extraction (APE) significantly reduced refining energy to a given freeness. For example, only 740 kWh/t specific refining energy was required to reach a freeness of 440 mL compared with 1340 kWh/t for the control.

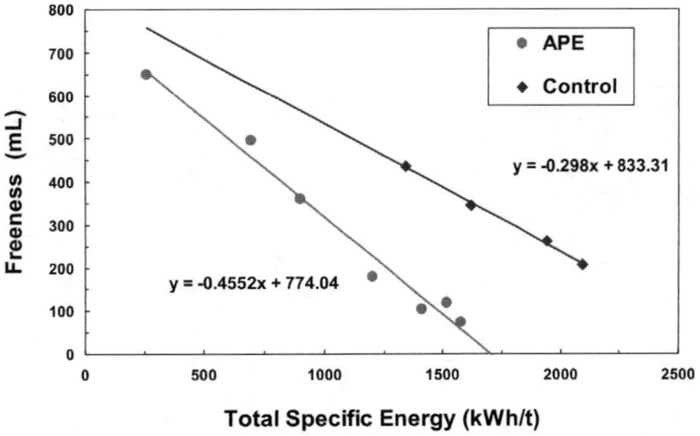

Figure 1. Acid pre-extraction significantly reduced the specific refining energy to a given freeness

Two pulps which had similar freeness around 350mL were selected for bleaching with 5% NaOH and 6.5% H_2O_2. Table 6 compared their physical properties before and after peroxide bleaching under the same conditions.

Table 6. Pulp physical properties before and after bleaching

Pulp Properties	Control	Acid pre-extracted
Unbleached:		
Refining energy, kWh/t	1640	910
Freeness, mL	344	361
Bulk, cm^3/g	3.28	2.88
Tensile index, N.m/g	13.1	11.4
Bleached:		
Freeness, mL	271	265
Bulk, cm^3/g	2.15	2.10
Tensile index, N.m/g	35.2	27.9
TEA, mJ/g	249	186
Gurley air resistance, s/100mL	8	7
PPS roughness, µm	5.7	5.8

Table 7 lists optical properties before and after bleaching for the same two pulps. With a peroxide charge of 6.5%, the control had an ISO brightness of 83% after bleaching compared to 79% for the acid pre-extracted pulp. The difference in brightness was still 4 points, the same difference as before bleaching. Table 7 also shows that the two bleached pulps had a similar light scattering coefficient but the light absorption coefficient of the control is lower than that of the acid pre-extracted pulp.

Table 7. Pulp optical properties before and after bleaching

Pulp Properties	Control	Acid pre-extracted
Unbleached:		
ISO brightness, %	57.7	53.5
Light scat. coeff., m2/kg	52.2	46.6
Light abs. coeff., m2/kg	2.75	3.42
b*	13.7	14.5
Bleached:		
ISO brightness, %	83.0	79.0
Light scat. coeff., m2/kg	39.8	38.2
Light abs. coeff., m2/kg	0.28	0.33
b*	5.6	8.3

4. Concluding Remarks

The benefits of the pre-extraction of aspen wood chips are:

- Recovering xylose and acetic acid
- Significantly reduces refining energy to a given freeness

Current drawbacks of prehydrolysis can be summarized as:

- Low hemicellulose extraction yield
- Lower brightness ceiling of 79% – can not be bleached to 85% brightness with peroxide
- Lower strength at a given freeness

Properties not affected:

- Bulk, surface roughness, and air resistance

This study will help to design better prehydrolysis processes while keeping pulp properties intact.

Acknowledgements

Funding from Industry Canada is greatly appreciated.

References

1. Zhuang., Q., and Vidal, P.F., Hemicellulose solubilisation from populus tremuloides via steam-explosion and characterization of the water soluble fraction., Cellulose Chem. Technol, 30, 371-384 (1996).
2. Teleman, A., Nordstrom, M., Tenkanen, M., Jacobs, A., Dahlman, O., Isolation and characterization of O-acetlylated glucomannans from aspen and birch wood, Carbohydrate Research 338, 525-534 (2003).
3. Gabrielli, I., Gatenholm, P., Glasser, W.G., Jain, R.K., Kenne, L., Seperation, characterization and hydro-gel formation of hemicellulose from aspen wood., Carbohydrate Polymers, 41, 1-7 (2000).
4. Fengel, D., and G. Wegener, Wood: Chemistry, Ultrastructure, Reactions, Walter de Gruyter, Berlin, (1989).

5. Pettersen, R. C., The Chemical Composition of Wood, in "Advance in Chemistry Series 207: The Chemistry of Solid Wood", Rowell, R., Ed., ASC, Washington, D. C., 57-126, (1984).
6. Sjostrom, E., Wood Chemistry, Fundamentals and Applications, Academic Press, New York (1981).
7. Gustavsson, M., Bengtsson, M., Gatenholm, P., Glasser, W., Teleman, A., and Dahlman, O., Isolation, Characterization and Material Properties of 4-O-methylglucuronoxylan, in "Biorelated Polymers: Sustainable Polymer Science and Technology, Ed. Chiellini et.al.", Kluwer Academic (2001).
8. Rowell, R., Lange, S., McSweeney, J., and Davis, M., Modification of wood fiber using steam, in "Proceeding The 6th Pacific Rim Bio_Based Composites Symposium", Volume 2, Wood Science and Engineering Department, Oregon State University, (2002).
9. Timell, T.E., Recent progress in the chemistry of wood hemicelluloses, in "Wood Science Technology", 1:45-70 (1967).
10. Lindquist, K., Simonson, R., Tingsvik, K., Studies on lignin carbohydrate linkages in milled wood lignin preparations, Svenks Papperstidning, 83:454-454 (1980).

199d Analysis of Transcriptional Response and Tolerance to Furfural in Industrial Yeast

Ying-jin Yuan, Department of Pharmaceutical Engineering, School of Chemical Engineering and Technology, Tianjin University, Tianjin, China

To overcome various bottlenecks and achieve key technological breakthroughs to provide technical support for the economic production of cellulose based biofuels. We employ systems biology tools to elucidate the microbial capability of growing and producing bioethanol under extreme or stressed conditions, to reveal various molecular and biochemical mechanisms underlying its tolerance. The response to furfural in exponential phase of both wild industrial budding yeast and the tolerant strain was characterized by measuring transcript levels of the genome on a global genomic cDNA microarray. It was found that 652 genes and 121 genes were regulated more than twofold in response to furfural in wild industrial budding yeast and the tolerant strain, respectively. Comparing the two strains, we found that the difference in energy metabolism and stress response might contribute to the furfural tolerance.

199e Catalytic Liquefaction of High-Pure Cellulose on ClO4-/ Fe2O3-ZrO2-La2O3

Xuewen Huang[1], **Zhiping Le**[1], Fei Yu[2], Zhaoping Chen[1], Lizhi Hong[1], and Yanqiu Huang[3]. (1) Chemistry, Nanchang University, 999 Xuefu Road, Qianhu, Nanchang 330031, China, (2) Dept. of Forest Products, Mississippi State University, 100 Blackjack Road, Starkville, MS 39762, (3) Graduate School, Nanchang University, 999 Xuefu Road, Qianhu, Nanchang, China

Biomass is the third largest source of energy for human being, and is also a renewable resource. High effective utilization and development of biomass energy has positive effects on solving energy and environment problems. Biomass contains 38%~50% cellulose. In this work, we prepared new kind of catalysts for the conversion of high-pure cellulose. The liquefaction of high-pure cellulose was carried out in a stainless autoclave, distilled water as liquefaction solvent, ClO4-/ Fe2O3- ZrO2 -La2O3 or ClO4-/Fe2O3 as catalyst. The results of liquefaction process showed that the ClO4-/ Fe2O3- ZrO2 -La2O3 catalyst has higher activity than ClO4-/Fe2O3. For example, under the reaction condition of 523 K, 3h, using ClO4-/ Fe2O3- ZrO2 -La2O3 as catalyst, the resident of high-pure cellulose was about 9.8%, but ClO4-/Fe2O3 was about 22.1%. In this test, the effect of the amount of the catalyst, reaction temperature and reaction time on the catalyst activity for the biomass conversion were investigated. This work is supported by grants from the Jiangxi Provincial Department of Education Fund (High Education & Technology) of China, the Jiangxi Provincial Department of Education Graduate Innovation Fund of China.

207a Electro-Oxidation of Ethanol Using Pt-Ru/C and Pt-Pd/C on Ballard Gas Diffusion Layers in PEM Fuel Cells

Virendra K. Mathur, Dept. of Chem. Eng., University of New Hampshire, 260 Kingsbury Hall, Durham, NH 03824 and Ryan S. Banfield, Dept. of Chem. Eng., University of New Hampshire, W311 Kingsbury Hall, Durham, NH 03824.

The direct ethanol fuel cells have attracted universal attention because ethanol has low volatility, no toxicity,inexpensive, high availability and easily transported. Our previous work on methanol with PtRu/C on Ballard gas diffusion layers has shown that superior DMFC performance can be obtained. The oxidation of ethanol is difficult and sluggish. Several catalysts coated on micro-porous Ballard gas diffusion layers are evaluated at various experimental conditions. Results will be presented in this paper.

Mathematical Formulation for the Performance Analysis of a Lithium-Ion Insertion Cell

S.S. Sandhu[*], *G.W. Brutchen*
Department of Chemical and Materials Engineering
University of Dayton
Dayton, OH 45469 USA

J.P. Fellner
Plan and Analysis Branch, Power Division, Propulsion Directorate
Air Force Research Laboratory
Wright-Patterson AFB, OH 45433-7251

[*]Corresponding author, email: Sarwan.Sandhu@notes.udayton.edu

Abstract

Understanding and analytical representation, through formulas, of the thermodynamic behavior, and transport and electrokinetic phenomena associated with a lithium-ion cell comprised of active, insertion material electrodes is of paramount importance for its design/performance evaluation. This paper presents a summary of the formulation to be used to predict the open-circuit electrode and total cell electric potentials, electrokinetic activation voltage losses or polarizations at the electrode-electrolyte interfaces, and temperature of a lithium-ion cell comprised of active-material, insertion electrodes.

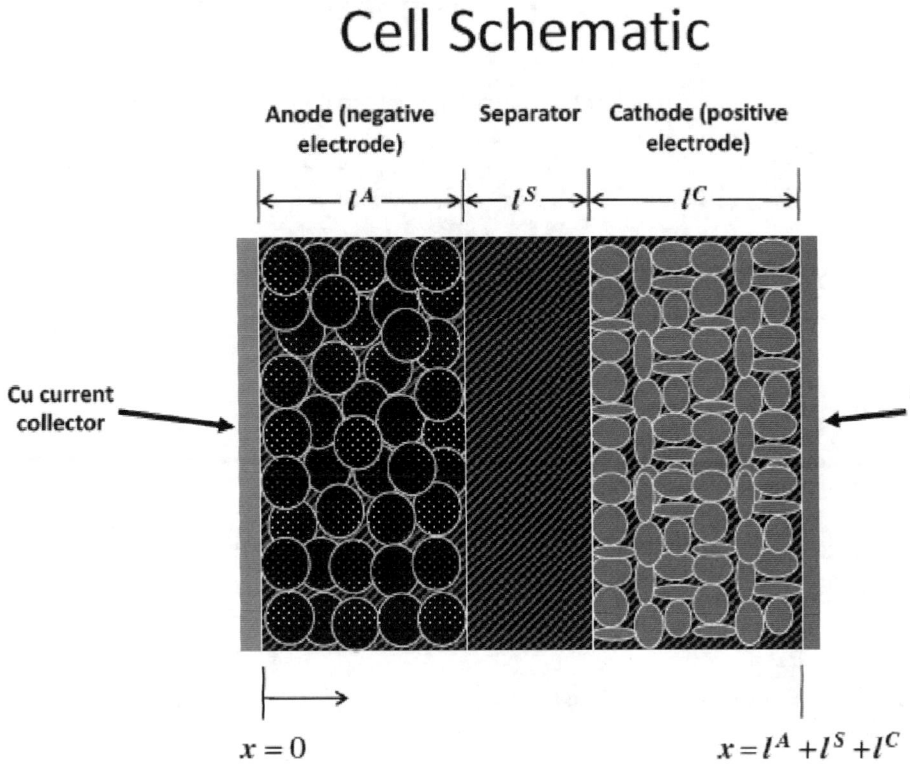

Figure 1

Cell Mathematical Formulation

Thermodynamic Formulation

Using principles of classical thermodynamics, the following equations have been developed to predict the cell anode and cathode open-circuit electric potentials.

For the anode the electric potential difference across the anode/solution interface, U^{A}, is given by:

$$\left(\Phi_1^{\mathrm{A}} - \Phi_2^{\mathrm{sol}}\right)_{\mathrm{rev}} = U^{\mathrm{A}} = U^{0,\mathrm{A}} + \left(\frac{RT}{F}\right)\ln\left(\frac{\gamma_{\mathrm{Li^+}}}{\gamma_{\mathrm{Li^+}}^{\mathrm{A}}}\right) + \left(\frac{RT}{F}\right)\ln\left(\frac{c/c_s^0}{x/x^0}\right) \quad (1)$$

where Φ_1^{A} is the electric potential of the bulk anode material, Φ_2^{sol} is the electric potential of the bulk electrolyte solution, $U^{0,\mathrm{A}}$ is the anode standard state potential, R is the universal gas constant, F is Faraday's constant, T is the temperature [K], $\gamma_{\mathrm{Li^+}}$ is the activity coefficient of Li^+ in the bulk electrolyte solution, $\gamma_{\mathrm{Li^+}}^{\mathrm{A}}$ is the activity coefficient of Li^+ in the anode solid phase, *c* is the electrolyte concentration, c_s^0 is the reference electrolyte concentration, *x* is the state of charge of the anode, and x^0 is the standard state-of-charge of the anode.

For the cathode the electric potential difference across the cathode/solution interface, U^{C}, is given by:

$$\left(\Phi_1^{\mathrm{C}} - \Phi_2^{\mathrm{sol}}\right)_{\mathrm{rev}} = U^{\mathrm{C}} = U^{0,\mathrm{C}} + \left(\frac{RT}{F}\right)\ln\left(\frac{\gamma_{\mathrm{Li^+}}}{\gamma_{\mathrm{Li^+}}^{\mathrm{C}}}\right) + \left(\frac{RT}{F}\right)\ln\left(\frac{c/c_s^0}{y/y^0}\right) \quad (2)$$

where Φ_1^{C} is the electric potential of the bulk cathode material, $U^{0,\mathrm{C}}$ is the cathode standard state potential, $\gamma_{\mathrm{Li^+}}^{\mathrm{C}}$ is the activity coefficient of Li^+ in the cathode solid phase, *y* is the state of charge of the cathode, and y^0 is the standard state-of-charge of the cathode.

The cell open-circuit electric potential is given by:

$$U^{\mathrm{cell}} = U^{\mathrm{C}} - U^{\mathrm{A}} = U^{0,\mathrm{cell}} + \left(\frac{RT}{F}\right)\ln\left(\frac{\gamma_{\mathrm{Li^+}}^{\mathrm{A}}}{\gamma_{\mathrm{Li^+}}^{\mathrm{C}}}\right) + \left(\frac{RT}{F}\right)\ln\left(\frac{x/x^0}{y/y^0}\right) \quad (3)$$

Electrode Electrochemical Kinetic Formulation

Using the basic principles of electrokinetics the following equations have been developed to compute the electrolyte solution-electrode interfacial surface current densities or activation voltage losses.

The cathode electrochemical reaction is:

$$\mathrm{Li}^{+}\begin{pmatrix}\text{in the}\\ \text{electrolyte}\\ \text{solution}\end{pmatrix} + \mathrm{e}^{-}\begin{pmatrix}\text{in the cathode}\\ \text{active material}\end{pmatrix} \underset{k_a^C}{\overset{k_c^C}{\rightleftarrows}} \mathrm{Li}\begin{pmatrix}\equiv\left(\mathrm{Li}^{+}+e^{-}\right)\text{ in the}\\ \text{cathode active material}\end{pmatrix} \tag{4}$$

and the net cathode current density, $i^{\mathrm{C}}_{\mathrm{n,\,surf}}$, is:

$$i^{\mathrm{C}}_{\mathrm{n,\,surf}} = i_0^{\mathrm{C}}\left[e^{\frac{\alpha_c^C F}{RT}\left(-\eta_s^C\right)} - e^{\frac{\alpha_a^C F}{RT}\left(\eta_s^C\right)}\right] \tag{5}$$

where i_0^{C} is the exchange current density for the cathodic reaction at the cathode-electrolyte interface, $\alpha_{\mathrm{a}}^{\mathrm{C}}$ and $\alpha_{\mathrm{c}}^{\mathrm{C}}$ are the anodic and cathodic direction reaction charge transfer coefficients for the cathode, respectively, and η_s^{C} is the activation overpotential for cathode reaction. Equation (6) shows how i_0^{C} is related to the prevailing salt concentration in the electrolyte, the cathode electrode state-of-charge (SOC), and the exchange current density at their reference state values at the prevailing temperature. Also, this assumes that the reaction activation energies are independent of *y* (SOC).

The cathodic exchange current density, i_0^{C}, varies with the electrolyte concentration and the cathode state-of-charge as shown by the following derived equation:

$$i_0^{\mathrm{C}} = i^{\mathrm{C}}_{0,\,\mathrm{ref}}\left(\frac{c}{c_{\mathrm{ref}}}\right)^{\alpha_{\mathrm{a}}^{\mathrm{C}}}\left(\frac{y}{y_{\mathrm{ref}}}\right)^{\alpha_{\mathrm{c}}^{\mathrm{C}}}\left(\frac{1-y}{1-y_{\mathrm{ref}}}\right)^{\alpha_{\mathrm{a}}^{\mathrm{C}}} \tag{6}$$

whe c and c_{ref} are the electrolyte concentration and the reference electrolyte concentration, respectively, and $i^{\mathrm{C}}_{0,\,\mathrm{ref}}$ is the exchange current density at the reference state values of c_{ref} and y_{ref}.

For a low current density situation $i^{\mathrm{C}}_{\mathrm{n,\,surf}}$ can be expressed as follows:

$$i^{\mathrm{C}}_{\mathrm{n,\,surf,\,low}} = \left[\left(\alpha_{\mathrm{a}}^{\mathrm{C}}+\alpha_{\mathrm{c}}^{\mathrm{C}}\right)\frac{F}{RT}\right]\left(i_0^{\mathrm{C}}\right)\left(-\eta_s^{C}\right) \tag{7}$$

and for a high current density situation $i^{\mathrm{C}}_{\mathrm{n,\,surf}}$ can be expressed as follows:

$$i^{\mathrm{C}}_{\mathrm{n,\,surf,\,high}} = i_0^{\mathrm{C}} e^{\frac{\alpha_c^C F}{RT}\left(-\eta_s^C\right)} \tag{8}$$

The anode electrochemical reaction is:

$$\mathrm{Li}\begin{pmatrix}\text{in the anode}\\ \text{active material}\end{pmatrix} \underset{k_c^A}{\overset{k_a^A}{\rightleftarrows}} \mathrm{Li}^{+}\begin{pmatrix}\text{in the}\\ \text{electrolyte}\\ \text{solution}\end{pmatrix} + \mathrm{e}^{-}\begin{pmatrix}\text{in the anode's}\\ \mathrm{e}^{-}\text{ conducting material}\end{pmatrix} \tag{9}$$

and the net anode current density, $i_{\text{n, surf}}^{\text{A}}$, is:

$$i_{\text{n, surf}}^{\text{A}} = i_0^{\text{A}} \left[e^{\frac{\alpha_{\text{a}}^{\text{A}} F}{RT}\left(\eta_{\text{s}}^{\text{A}}\right)} - e^{\frac{\alpha_{\text{c}}^{\text{A}} F}{RT}\left(-\eta_{\text{s}}^{\text{A}}\right)} \right] \tag{10}$$

where i_0^{A} is the exchange current density for the anodic reaction at the anode-electrolyte interface, where $\alpha_{\text{a}}^{\text{A}}$ and $\alpha_{\text{c}}^{\text{A}}$ are the anodic and cathodic direction reaction charge transfer coefficients for the anode, respectively and $\eta_{\text{s}}^{\text{A}}$ is the activation overpotential for anode reaction. Equation (11) shows how i_0^{A} is related to the prevailing salt concentration in the electrolyte, the anode electrode state-of-charge (SOC), and the exchange current density at their reference state values at the prevailing temperature. Also, this assumes that the reaction activation energies are independent of *x* (SOC).

The anodic exchange current density, i_0^{A}, varies with the electrolyte concentration and the anode state-of-charge as shown by the following derived equation:

$$i_0^{\text{A}} = i_{0,\text{ ref}}^{\text{A}} \left(\frac{c}{c_{\text{ref}}}\right)^{\alpha_{\text{a}}^{\text{A}}} \left(\frac{x}{x_{\text{ref}}}\right)^{\alpha_{\text{c}}^{\text{A}}} \left(\frac{1-x}{1-x_{\text{ref}}}\right)^{\alpha_{\text{a}}^{\text{A}}} \tag{11}$$

For a low current density situation $i_{\text{n, surf}}^{\text{A}}$ can be expressed as follows:

$$i_{\text{n, surf, low}}^{\text{A}} = \left[\left(\alpha_{\text{a}}^{\text{A}} + \alpha_{\text{c}}^{\text{A}}\right)\frac{F}{RT}\right]\left(i_0^{\text{A}}\right)\left(\eta_{\text{s}}^{A}\right) \tag{12}$$

and for a high current density situation $i_{\text{n, surf}}^{\text{A}}$ can be expressed as follows:

$$i_{\text{n, surf, high}}^{\text{A}} = i_0^{\text{A}} e^{\frac{\alpha_{\text{a}}^{\text{A}} F}{RT}\left(\eta_{\text{s}}^{\text{A}}\right)} \tag{13}$$

<u>Transient Thermal Energy Balance</u>

By the application of the first law of thermodynamics to a lithium-ion insertion cell, the following equation was obtained.

$$C_p \frac{dT}{dt} = \left(\text{U}^{\text{cell}} - V^{\text{cell}}\right)\text{I} + \left[T\left(\text{-}\Delta\text{S}\right)\frac{\text{I}}{nF}\right] + h_0\left(T_a - T\right) \tag{14}$$

where C_p is the cell heat capacity, V^{cell} is cell voltage under load, I is the geometric current density based upon the unit geometric area of the separator, ΔS is the change in entropy associated with the occurrence of one gram-mole of the cell reaction, *n* is the gram-moles of electrons involved in the cell reaction, h_0 is the heat transfer coefficient for heat transfer between the cell and its surroundings, and T_a is the cell environmental temperature [K].

Now, since $\left(\frac{\Delta S}{nF}\right)=\left(\frac{\partial U^{cell}}{\partial T}\right)_{P,SOC}$, Eq. (14) can be written:

$$C_p \frac{dT}{dt}=\left[U^{cell}-T\left(\frac{\partial U^{cell}}{\partial T}\right)_{P,SOC}-V^{cell}\right]I+h_0\left(T_a-T\right) \tag{15}$$

Rearranging and incorporating the standard-state open circuit or reversible electric potential of the cell, $U^{0,cell}$, at a given temperature, T [K] Eq. (15) becomes:

$$\frac{dT}{dt}=\frac{\left[U^{0,cell}-V^{cell}-T\frac{dU^{0,cell}}{dT}\right]I+h_0\left(T_a-T\right)}{C_p} \tag{16}$$

It is here noted that the derivation of Eq. (16) is based on the assumption of negligible variation of the ratio of lithium ion activity coeffficient in the anode insertion material to that in the cathode insertion material with respect to temperature.

In the case where $\frac{dT}{dt}=0$ we can rearrange Eq. (16) to determine the steady-state temperature, T_{ss} [K], of the cell:

$$T_{ss}=\frac{\left[\left(IU^{0,cell}\right)-\left(IV^{cell}\right)\right]+h_0T_a}{\left(\frac{dU^{0,cell}}{dT}\right)I+h_0} \tag{17}$$

Application of the Developed Formulation

To illustrate the application of the developed thermodynamic formulation, the open-circuit experimental data on the nickel dioxide and disordered carbon anode [8] were employed. The following polynomials were developed using a software package, Polymath 5.1, to compute the activity coefficients of lithium ion in the cathode nickel dioxide active material and anode carbon active material.

$$\ln\gamma_{Li^+}^{C}=a_0^C+a_1^C\left(\frac{y}{y^\circ}\right)+a_2^C\left(\frac{y}{y^\circ}\right)^2+a_3^C\left(\frac{y}{y^\circ}\right)^3+a_4^C\left(\frac{y}{y^\circ}\right)^4+a_5^C\left(\frac{y}{y^\circ}\right)^5$$
$$\left(valid\ for\ 1\le\frac{y}{y^\circ}\le 2.1111;\ y^\circ=0.45\right) \tag{18}$$

$$\ln\gamma_{Li^+}^{A}=a_0^A+a_1^A\left(\frac{x}{x^\circ}\right)+a_2^A\left(\frac{x}{x^\circ}\right)^2+a_3^A\left(\frac{x}{x^\circ}\right)^3+a_4^A\left(\frac{x}{x^\circ}\right)^4+a_5^A\left(\frac{x}{x^\circ}\right)^5$$
$$\left(valid\ for\ 1\le\frac{x}{x^\circ}\le 14;\ x^\circ=0.05\right) \tag{19}$$

The regression analysis generated coefficients for use in Eqs. (18) and (19) are:

$a_0^C = -512.6349$, $a_1^C = 1564.5991$, $a_2^C = -1922.8342$, $a_3^C = 1197.8539$,
$a_4^C = -373.38256$, $a_5^C = 46.39578$

Correlation coefficient, $\left(R^C\right)^2 = 1$

$a_0^A = -7.7967566$, $a_1^A = 8.3373677$, $a_2^A = -0.5581383$, $a_3^A = 0.0112673$, $a_4^A = 5.804 \text{ x } 10^{-4}$,
$a_5^A = -2.622 \text{ x } 10^{-5}$

Correlation coefficient, $\left(R^A\right)^2 = 0.999999.$

The electrode kinetics formulation was used to compute the activation polarization voltage losses at the cathode and anode electrodes for a $Li_xC_6 / Li_yMn_2O_4$ cell, with $LiPF_6$ in an organic solvent mixture held in *p(Vdf-HFP)* as an electrolyte, at 25 °C.

- At low discharge rate, e.g.; 1/10 C-rate ≡ 1.75 x 10^{-4} A cm_{geom}^{-2}; therefore low η_S values:

$$\left(-\eta_S^C\right) \text{ at cathode} = 3.1 \text{ mV}; \ \left(\eta_S^A\right) \text{ at anode} = 4.0 \text{ mV}$$

- At high discharge rate, e.g.; 100 C-rate ≡ 1.75 x 10^{-1} A cm_{geom}^{-2}; therefore, high η_S values:

$$\left(-\eta_S^C\right) \text{ at cathode} = 245 \text{ mV}; \ \left(\eta_S^A\right) \text{ at anode} = 259 \text{ mV}$$

Data used: Anode active material particle radius, R_S^A = 12.5 μm; cathode active material particle radius, R_S^C = 8.5 μm; electrode thicknesses, $\delta^A = 100\mu m$, $\delta^C = 175\mu m$; volume fractions occupied by the cell electrolyte (l), polymer (p), and filler (f) in the electrodes: $\varepsilon_l^A = 0.357$, $\varepsilon_p^A = 0.146$, $\varepsilon_f^A = 0.026$, $\varepsilon_l^C = 0.444$, $\varepsilon_p^C = 0.186$, $\varepsilon_f^C = 0.073$. The exchange current densities, $i_\circ^A = 10^{-4} A\, cm_{surf}^{-2}$, $i_\circ^C = 8 \text{ x } 10^{-5} A\, cm_{surf}^{-2}$.

The thermal energy balance Eq. (15) was applied to the lithium-ion cell during its discharge. The temperature versus time profiles are shown in Fig. 2 for the geometric current densities, *I* = 1.75 $\frac{mA}{cm_{geom}^2}$ (≡ 1 C-rate), *I* = 3.5 $\frac{mA}{cm_{geom}^2}$ (≡ 2 C-rate), *I* = 5.25 $\frac{mA}{cm_{geom}^2}$ (≡ 3 C-rate) and *I* = 7.0 $\frac{mA}{cm_{geom}^2}$ ≡ 4 C-rate).

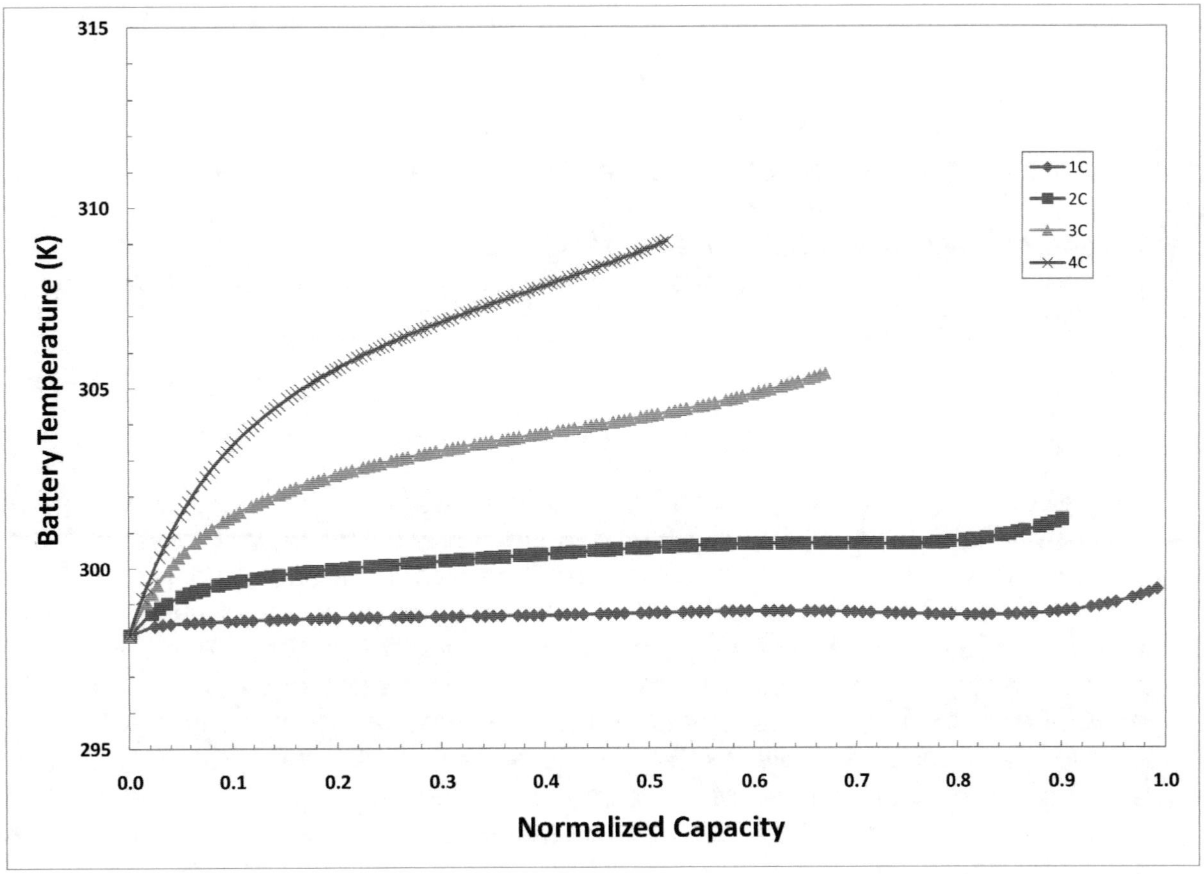

Figure 2. Cell Temperature versus Normalized Capacity at Different Discharge Rates.

<u>Concluding Remarks</u>

Formulation to predict the open-circuit electric potential, activation polarization voltage losses at the electrolyte-insertion material interface in the cell electrodes, and cell temperature has been developed for a typical lithium insertion electrode cell as shown in Fig. 1.

Application of the developed formulation has been illustrated. The future work has been planned to predict the cell voltage of a lithium insertion cell based on the presented formulation.

Modeling High Rate Lithium-Ion Batteries Using Fundamental and Empirical Methods

Paper Number 207c

Joseph P. Fellner and Gary J. Loeber, Air Force Research Laboratory, Wright-Patterson AFB, OH, USA
Sarwan S. Sandhu, University of Dayton, Dayton, OH, USA
Binod Kumar, Joykumar S. Thokchom, and Cameron Riepenhoff, University of Dayton Research Institute, Dayton, OH, USA

Abstract

The power density of high rate lithium-ion batteries and lithium-ion polymer batteries is increasing yearly. To determine the viability of these batteries for powering various different types of electrically-powered devices, their performance under various conditions needs to be predicted/determined. We have measured the performance of commercially available high power cells from A123 under various temperatures and charge/discharge conditions. We have also used AC impedance spectroscopy to determine the equivalent circuits of these cells as a function of temperature and charge state. From the impedance measurements, we have developed empirically based models for these cells and the fidelity of these empirical models will be presented. Also, with the use of a commercial battery modeling software program, we have initiated the modeling of these batteries from a fundamental basis and the fidelity of these models will also be presented.

Introduction

Commercially available, high power density lithium-ion batteries are currently being used for powering electric drills and remotely-controlled aircraft. Performance of these types of batteries as a function of temperature, cycle number, and state-of-charge will help determine what applications these batteries can be used in. Verified models that accurately simulate the performance of the types of cells will also aid in optimizing application selection.

Lithium-Ion Cell Performance

Commercially available high power density lithium-ion batteries from various companies were procured for performance characterization. See Figure 1 for the cell specification from A123's website. The capacities of the A123 cell as a function of discharge rates are given in Figure 2. Figure 3 shows a typical charge curve for the A123 cell. Note from Figure 2 that the A123 cell can be discharged at 20 times its nominal 1 hour discharge rate (20C-rate) and still retains over 90% of its coulombic capacity. Also note the temperature rise for the cell during the high discharge rates.

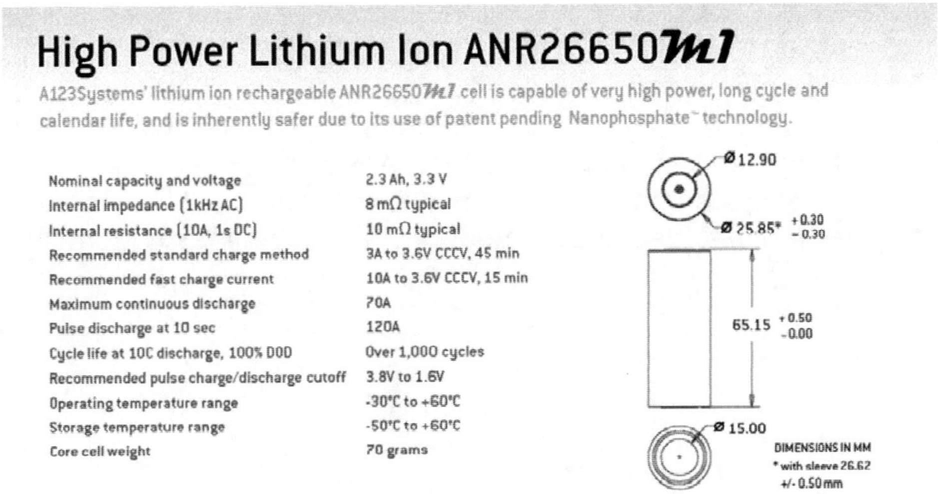
High Power Lithium Ion ANR26650*m1*

A123Systems' lithium ion rechargeable ANR26650*m1* cell is capable of very high power, long cycle and calendar life, and is inherently safer due to its use of patent pending Nanophosphate™ technology.

Nominal capacity and voltage	2.3 Ah, 3.3 V
Internal impedance (1kHz AC)	8 mΩ typical
Internal resistance (10A, 1s DC)	10 mΩ typical
Recommended standard charge method	3A to 3.6V CCCV, 45 min
Recommended fast charge current	10A to 3.6V CCCV, 15 min
Maximum continuous discharge	70A
Pulse discharge at 10 sec	120A
Cycle life at 10C discharge, 100% DOD	Over 1,000 cycles
Recommended pulse charge/discharge cutoff	3.8V to 1.6V
Operating temperature range	-30°C to +60°C
Storage temperature range	-50°C to +60°C
Core cell weight	70 grams

Figure 1. A123 26650 Cell Specifications

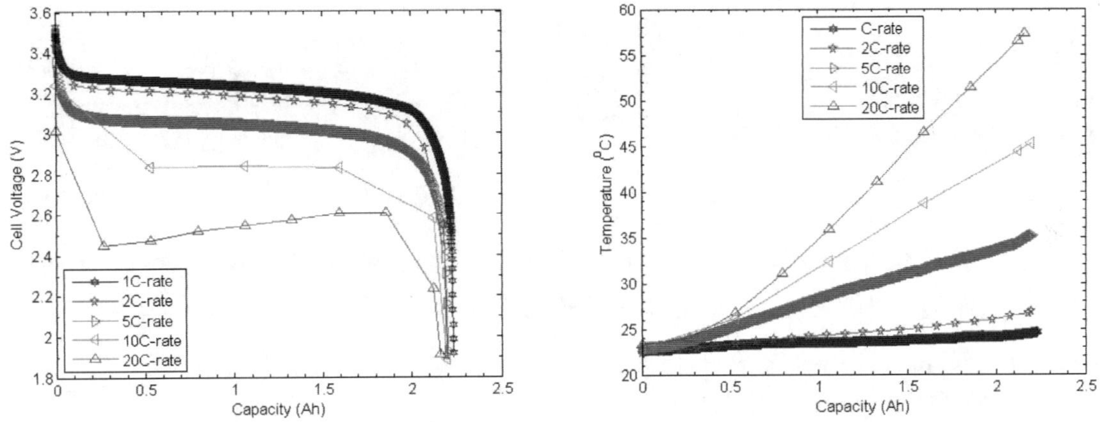

Figure 2. A123 26650 Cell Voltage and Temperature as a Function of Discharge Rate

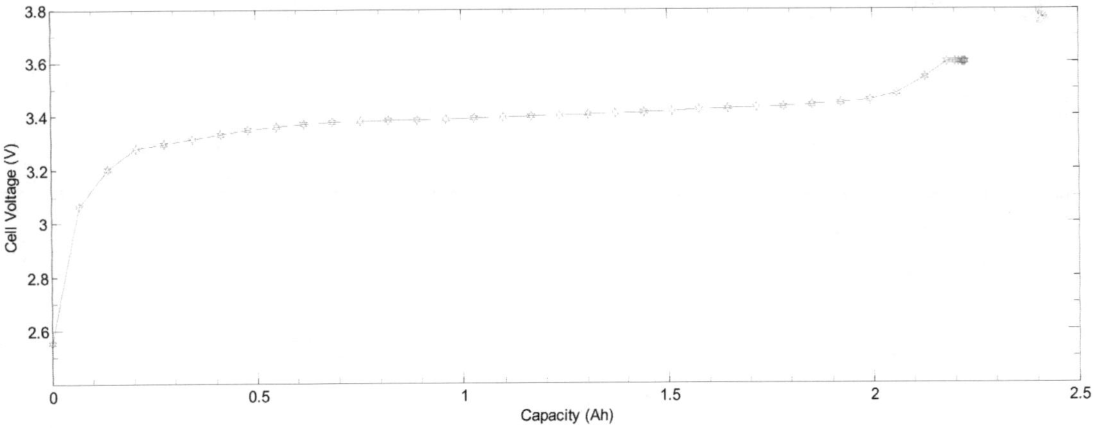

Figure 3. A123 26650 Cell Charged at 3 Amps and at Room Temperature

AC Impedance Spectroscopy

AC impedance spectroscopy was used to determine the cell Nyquist performance as a function of state-of-charge and temperature. Figure 4 show the Nyquist plots for an A123 26650 cell at 20% state-of-charge and different temperatures. The interfacial or charge transfer resistance was found to decrease exponentially with decreasing temperature.

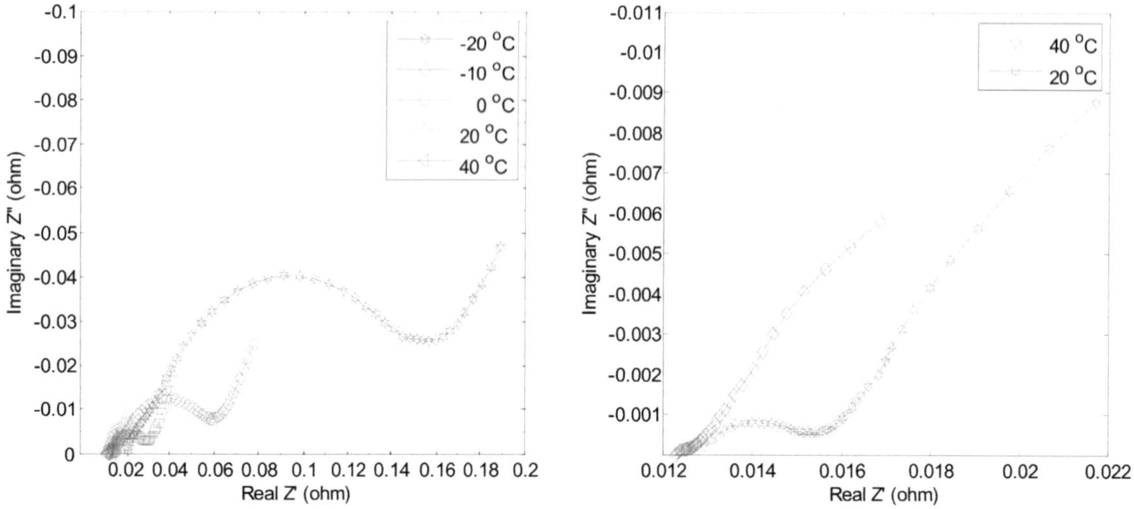

Figure 4. Nyquist Plot for an A123 26650 Cell at 20% State-of-Charge

From Figures 5 and 6, it clearly can be seen that cell impedances are not a function of the state-of-charge at a given temperature.

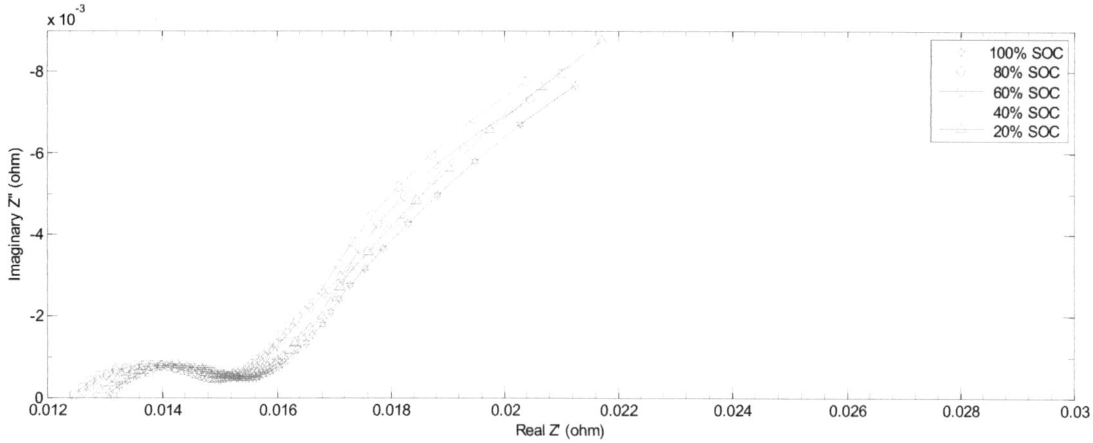

Figure 5. Nyquist Plot for an A123 26650 Cell at 20 °C and Various States-of-Charge

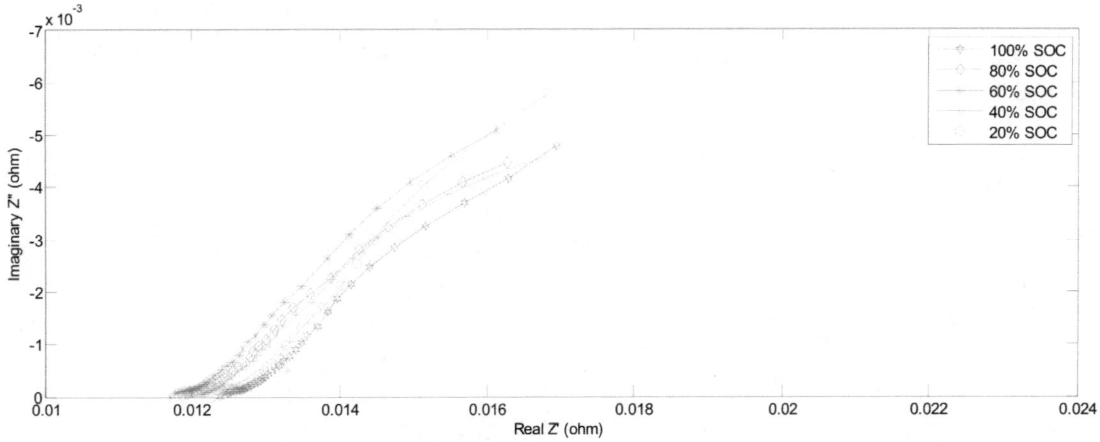

Figure 6. Nyquist Plot for an A123 26650 Cell at 40 °C and Various States-of-Charge

Equivalent Circuit Modeling

From the Nyquist plots for the A123 cell determined using AC impedance spectroscopy, simple equivalent circuits were developed. Figure 8 shows one based on two RC's, a series resistor, and an open-circuit voltage source whose voltage varies with state-of-charge.

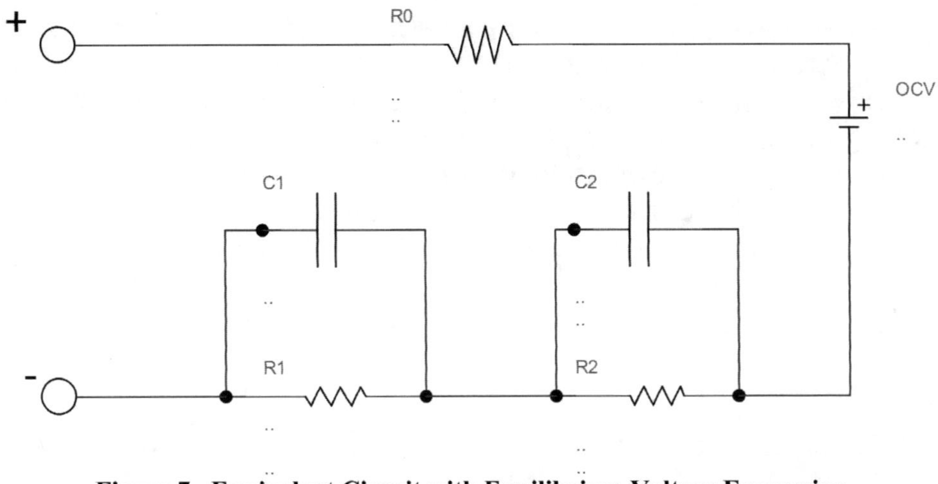

Figure 7. Equivalent Circuit with Equilibrium Voltage Expression

The open-circuit voltage expression was based on the Nernst equation and open-circuit cell measurements. The equation for the A123 open-circuit voltage is:

$$E = 3.2952 - \frac{RT}{F}\ln\left(\frac{1-SOC}{SOC}\right) \quad (1)$$

where *SOC* is the cell state-of-charge. The values for R1, R2, C1, and C2 were determined from the Nyquist plots. The value for R0 used was a reduced value from the Nyquist plots.

Figure 8 shows actual cell data and simulated data for a given high-rate discharge profile. The only parameter in the equivalent circuit that was changing as a function of time was the open-circuit voltage.

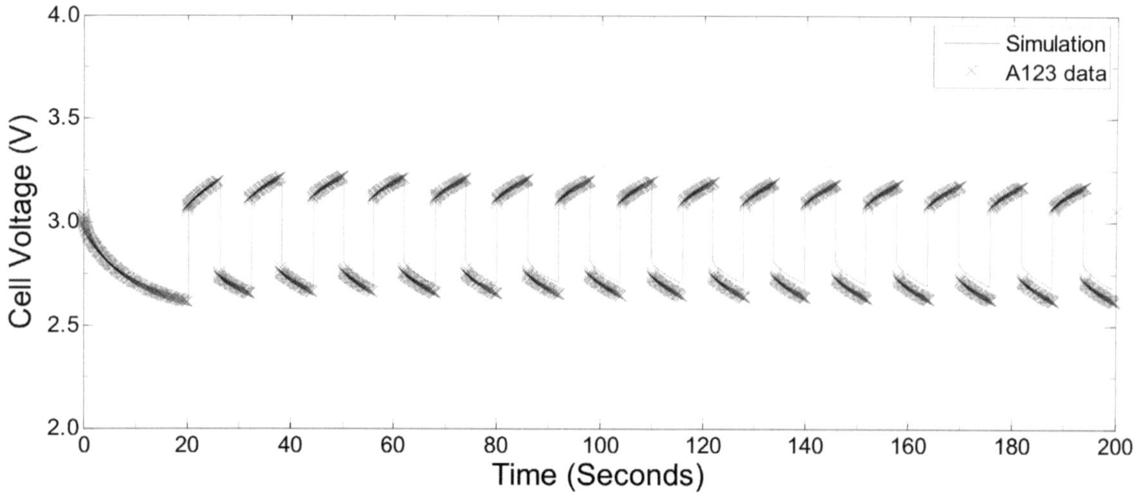

Figure 8. Simple Equivalent Circuit Simulation Versus Actual A123 26650 Cell Data at 46 Amp Current Pulses.

Summary/Conclusions

The performance of a commercially available cell was determined as a function discharge rate. Further cell characterization was performed using AC impedance spectroscopy. Cell impedance, in terms of cell interface or charge transfer resistance, was found to be a strong function of temperature. The Nyquist plots, indicative of cell impedance as a function of frequency/time, were relatively insensitive to state-of-charge for a given cell temperature. A simple equivalent circuit for the A123 cell was derived from the cell Nyquist plots and was used to accurately simulate the cell performance for a high current pulse discharge test.

Preparation and Characterization of higher loading Pt-Ru/C Catalysts

Shu-bo Wang[1], Xiao-feng Xie*[1], Zhi-kuan Yang[1], Yao-wu Wang[1], Pei-gang Jiang[2], Yong-an Lei[2], Virendra K. Mathur[3]

1. Institute of Nuclear and New Energy, Tsinghua University, Beijing 100084, China,
2. Beijing Beifen-Ruili Analytical Instrument (Group) Co., Ltd. Beijing 100095, China,
3. Dept. of Chem. Eng, University of New Hampshire, Durham, NH 03824

Corresponding author Tel: +86 10 6278 4827; fax: +86 10 6277 1150
Email address: xiexf@tsinghua.edu.cn (Xiao-feng Xie)

Abstract

A 60% PtRu/C anode electrocatalyst for a direct methanol fuel cell (DMFC) was prepared by impregnation method, using sodium hypophosphite as reductive agent. Effects of support on the catalyst activities were studied. The structure, morphology and electrocatalytic performances were characterized by XRD, XPS, and Linear Sweep Voltammetry (LSV). These results indicated that the structure and electrocatalytic properties were greatly influenced by the pH value of impregnating solution. The catalyst was prepared at the pH value of 7. It has higher dispersion, higher catalytic activity for methanol oxidation, good tolerance to poisoning species and the average particle size of 2.8nm. Compared with the previous PtRu/C catalyst, the PtRu/C catalyst possesses a higher catalytic activity producing current density of 60 $mAmg^{-1}$ at 0.45V (vs.SCE) with a scanning rate of 20mV/s.

Keywords: Direct methanol fuel cell; PtRu/C catalysts; Electrocatalysis

1. Introduction

It has been suggested that a bimetallic PtRu/C catalyst can perform better due to electronic effects. The PtRu alloy structure contributes little to the performance of the catalysts. The performance of bimetallic PtRu/C catalysts is dependent on a specific process, size, surface structure, morphology, etc. Lin et al experimentally demonstrated the different roles of citrate as a reluctant or a stabilizer and that the reducing reaction could not take place below 90 ◦C. Unfortunately, not many studies have been reported in carrier effects. On the other hand, it has been known for sometime that carrier influences the chemical equilibrium of weak carbonyl acid with Pt nanoparticles. In our previous study, citrate-stabilized PtRu/C anode catalysts were prepared systematically. Annealing treatment and stabilizing pH were separately controlled to develop homogeneous dispersed catalyst with adequate hydrous ruthenium oxide to promote the catalyst activity. In this study, a 60% PtRu/C anode electrocatalyst was prepared by impregnation method, using sodium hypophosphite as reductive agent. Effects of support on the catalyst activities were studied. The structure, morphology and electrocatalytic performances were characterized by XRD, XPS, and LSV. These results indicated that the structure and electrocatalytic properties were greatly influenced by the pH value of impregnating solution. Compared with the previous PtRu/C catalyst, the PtRu/C catalyst possesses a higher catalytic activity producing current of 5.7 mA at 0.45V (vs.SCE) with a scanning rate of 20mV/s.

2. Experiment

2.1. Preparation of catalysts

All chemicals used were analytical grade. A 60 wt% PtRu/C catalyst with a Pt to Ru molar ratio of 1:1 was prepared by impregnation method. Vulcan carbon powder XC-72, XC-72R and BP 2000 with a specific BET area of $250m^2g^{-1}$, $254m^2g^{-1}$ and $1475m^2g^{-1}$ respectively was purchased and used as supports. The supports were pretreated for 2 h at 400 ◦C in nitrogen atmosphere was dispersed in deionized water- called as A. H_2PtCl_6 and $RuCl_3$ solutions were added into A. The suspension A was ultrasonicated for 30 min. and impregnated another 16 h. An excess of NaH_2PO_2 solution was added to the ink as a reducing agent. The temperature was maintained at 90 ◦C for another 4 h to ensure a complete reduction. The prepared catalysts were filtered, washed thoroughly and dried in an air oven. The detail procedure is shown as fig1.

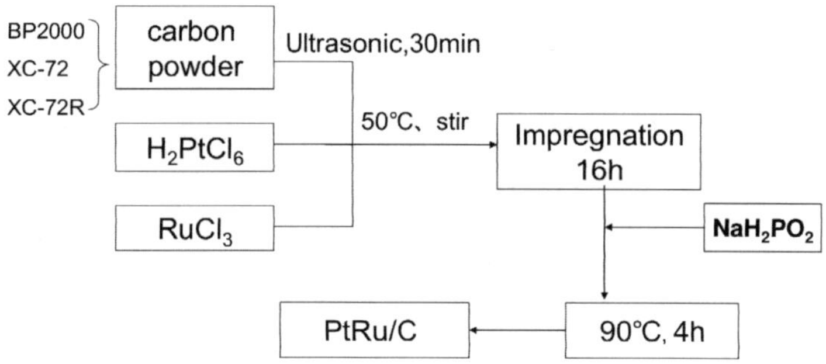

Fig.1 Catalyst preparation procedure delineation

2.2. Properties characterizations

Powder X-ray diffraction patterns were obtained using a Bruker D8-Discover micro X-ray diffractometer with a Cu Kα radiation source at 40 kV and 40 mA. The particle size of the catalysts was characterized by XRD. The X-ray photoelectron spectrum (XPS) was obtained with a PHI Quantera SXM spectrometer using Al monochromatic X-ray source operating at 4.0 kV. Observed spectra were deconvoluted by the Peak fit software fitted with the Gaussian function. Electrochemical measurements were carried out by a three electrode cell with a saturated calomel electrode (SCE) as the reference electrode, Pt foil as the counter electrode, and the prepared catalyst as the working electrode at ambient temperature. The three electrode cell system was purged with nitrogen gas for 30 min before the electrochemical analysis. The LSV curve was recorded at $5mVs^{-1}$ in the solution of 0.5M H_2SO_4 and 0.5M CH_3OH.

3. Results and discussions

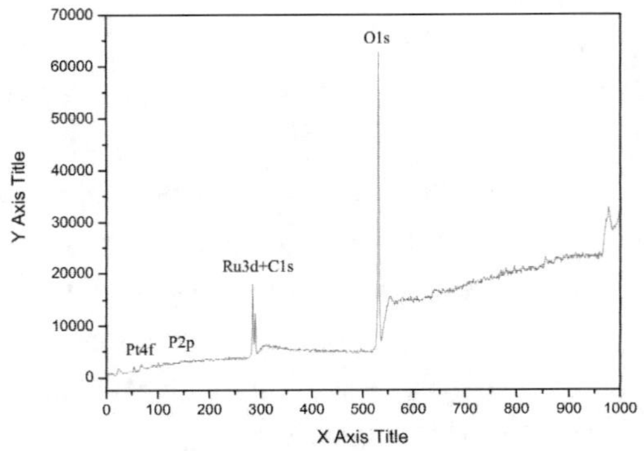

Fig. 2 XPS analysis of PtRu/C

The XPS spectrum of the catalysts is shown in Fig. 2. The results obtained by the deconvolution of the XPS spectra are summarized in Table 1. The most intensive peak of 284.75 eV was assigned to carbon graphite C1s, and the other two peaks are both attributed to carbon bonded to oxygen, C=O and COO−, respectively. The Pt4f spectrum consists of three couples of doublets with low binding energies (BE) of 71.72 eV, 72.65 eV, and 74.48 eV. The most intense doublet of 71.29 eV is attributed to a zero-valence metallic platinum which is shifted a little to a higher BE than that of the bulk Pt element (71.29 eV). This is because the interactions between Pt nanoparticles and carbon support. The doublet of 72.65 eV indicates the presence of amorphous Pt (II) species, and the broad doublet of 74.48 eV is assigned to the Pt (IV) species.

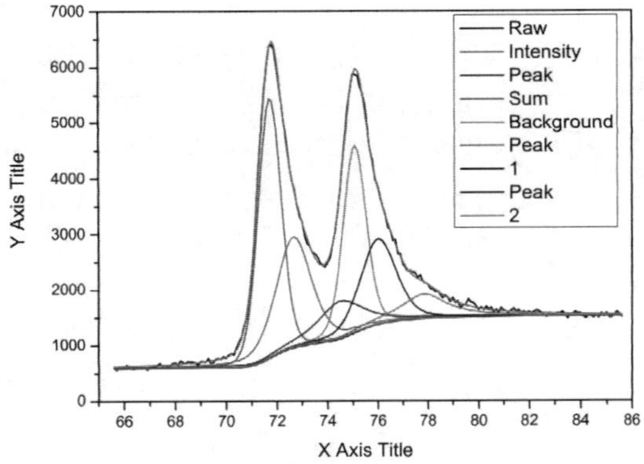

Fig. 3 Regional XPS of PtRu/C nanocatalyst in Pt4f

Table 1 Binding Energies and surface compositions

Species	Binding energy (eV)	Possible chemical state	Relative concentrations (%)
Pt4f	71.72	Pt^0	47.45
	72.65	Pt(II)	35.63
	74.48	Pt(IV)	16.91
Ru3d	281.93	Ru(IV)	100
C1s	284.75	C	84.81
	286.44	C=O	9.89
	288.75	COO-	5.29

The XRD pattern of the prepared catalyst was shown in Fig. 4. The catalyst XRD curve has noticeable P diffraction peaks. There are three peaks of metal Pt face-centered cubic (fcc) diffraction—Pt(1 1 1), Pt(2 2 0) and Pt(3 1 1) and are a little shifted to a higher 2θ values from 39.64◦, 66.85◦ and 80.86◦ for pure Pt fcc indicating the reduction in lattice constants because of Ru substitution. Moreover, no recognizable metallic ruthenium diffraction was observed. It can be said that the ruthenium may not only enter Pt fcc-lattice to establish PtRu alloy. The lattice parameters, for PtRu alloy in Pt (2 2 0) were 33.6%, 35.3% and 35.8% for the catalysts BP 2000, XC-72 and XC-72R, respectively by Vegard's law. It showed that the support did not significantly affect the degree of alloying. According to Scherer equation, the catalyst has the smallest size of 2.8 nm.

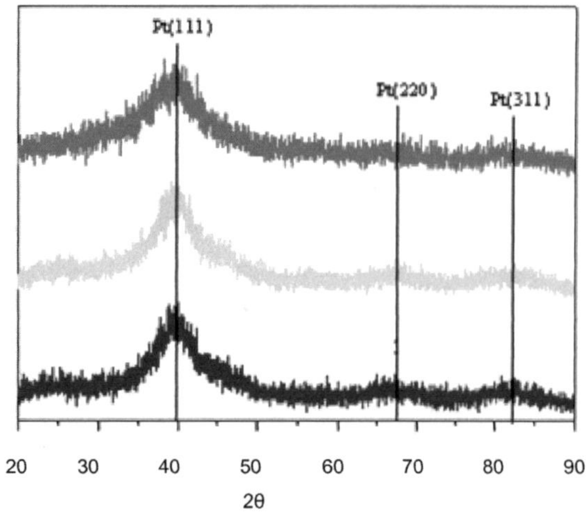

Fig. 4 XRD patterns of 60% PtRu/C catalysts

The LSV curves of methanol electro-oxidation on the catalysts are shown as Fig. 5. Each of the curves is the average of three parallel measurements. For avoiding anodic Ru electrooxidation, the LSV upper potential was fixed as 0.45V and dissoluted at a higher potential. The figure shows that the BP2000 catalyst has the highest current density ($60 mAmg^{-1}$) at the potential of 0.45V. This result indicated that the support BP2000 catalyst exhibits the best electrocatalytic activity for methanol electrooxidation. The surface area (BET-N_2) of both XC-72 and XC-72R is 254 m^2g^{-1}, while BP 2000 is about 1475 m^2g^{-1}. The greater the surface area is, the more the metal particles loaded. This increased active catalytic area to make the catalyst having more electrocatalytic activates.

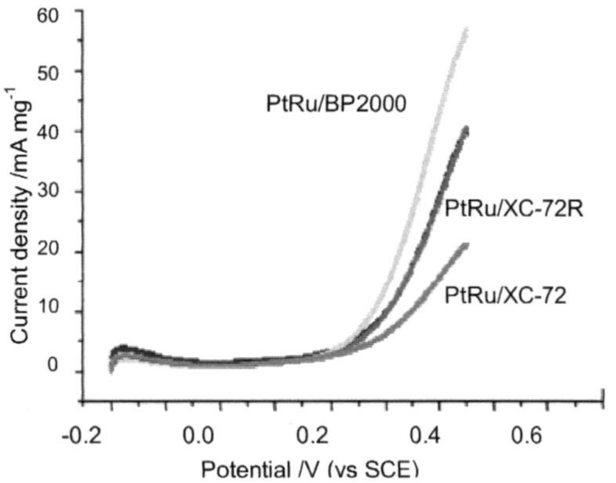

Fig. 5 LSV of methanol oxidation

The TG-DSC curve of PtRu/XC-BP2000 is shown as Fig. 6. After 200℃, there is a endothermic peak, it might be a PtRu alloying phase transition. The higher the temperature, the higher alloying degree.

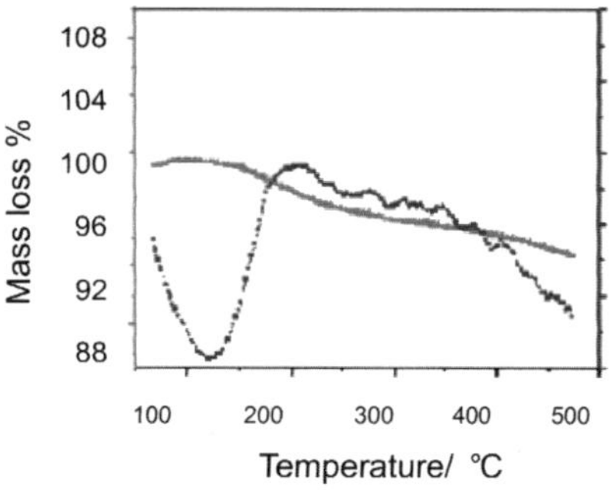

Fig. 6 TG-DTA curve of PtRu/BP2000

4. Conclusion

Higher loading PtRu/C catalysts were prepared by an impregnation method by NaH_2PO_2 as a reducing agent in an aqueous solution. Catalyst characteristics were studied using various supports. The results showed that the particle size could be controlled below 2 nm by changing operation factors. BP2000 had the best electrocatalytic because of its higher surface area among the supports. PtRu The higher the temperature, the higher the PtRu/C alloying degree. Further, a single cell DMFC performance measurement will make to demonstrate the electrocatalytic activities of the prepared catalyst.

Acknowledgements

The work was funded by the National High Technology R&D Program of China (2006AA05Z138) and Natural Science Foundation of China (50573041, 50703021).

References

[1] Y.Q. Xu, X.X. Xie, J.W. Guo, Sh.B. Wang, Y.W. Wang, V.K. Mathur, J. Power Sources 162 (2006) 132-140.

[2] Q. Ge, S. Desai, M. Neurock, K. Kourtakis, J. Phys. Chem. B 105 (2001)9533–9536.

[3] C. Bock, M.A. Blakely, B. MacDougall, Electrochim. Acta 50 (2005)2401–2414.

[4] J. Turkevich, R.S. Miner Jr., L. Babenkova, J. Phys. Chem. 90 (1986) 4765–4767.

[5] T. Iwasita, H. Hoster, A.J. Anacker, W.F. Lin, W. Vielstich, Langmuir 16(2000) 522–529.

[6] C.S. Lin, M.R. Khan, S.D. Lin, J. Colloid Interface Sci. 287 (2005)366–369.

[7] V. Radmilovic, H.A. Gasteiger, P.N. Ross Jr., J. Catal. 154 (1995) 98–106.

[8] P.V. Samant, C.M. Rangel, M.H. Romerob, J.B. Fernandes, J.L. Figueiredo,J. Power Sources 151 (2005) 79–84.

[9] S. Rojas, F.J. Garcıa-Garcıa, S. Jaras, M.V. Martınez-Huerta, J.L.G. Fierro,M. Boutonnet, Appl. Catal. A 285 (2005) 24–35.

207e Preparation of Thin YSZ Electrolyte by Electrostatic Spray Deposition for Solid Oxide Fuel Cells

Jinsoo Kim, Hyuk Taek Kwon, and Byeongwan Kwon. College of Environment and Applied Chemistry, Kyung Hee University, 1 Seochen-dong Giheung-gu, Yongin, 449-701, South Korea

Solid oxide fuel cell is one of the most competitive candidates for energy generation. However, it requires high operating temperature (> 800°) to obtain proper ionic conductivity of the YSZ electrolyte. The operating temperature can be lowered by reducing the thickness of the electrolyte. In this research, we prepared thin YSZ electrolyte by ESD technique. Electrostatic spray showed a few modes such as dripping, con-jet and multi-jet mode, depending on the applied voltage. SEM image confirmed that dense YSZ layer (10~20 μ) was formed on NiO-YSZ anode after annealing at 1500° for 3h. A single cell was prepared by screen printing LSCF layer on the electrospray-deposited YSZ/NiO-YSZ sample. The cell test was performed in the temperature range of 500~800°.

207f Cellulose Triacetate : A New Material for Polymer Gel Electrolyte

Dinh Quan Nguyen[1], Ji-hee Oh[1], Hyunjoo Lee[1], Hoon Sik Kim[2], and Sang Deuk Lee[1]. (1) Energy & Environment Research Division, Korea Institute of Science & Technology, 39-1, Hawolgok-dong, Seongbuk-gu, Seoul, South Korea, (2) Dept. of Chemistry, Kyung Hee University, 1 Hoegidong, Dongdaemoongu, Seoul 130-701, South Korea

Several gel electrolytes were prepared from 1-methyl-1-propylpyrrolidinium bis (trifluoromethanesulfonyl)imide and cellulose triacetate according to the molar ratios between cellulose triacetate, ionic liquid and LiTFSI. The optimal gel ingredient was recognized by measuring the unique melting point with DSC and optically confirmed by SEM in order to obtain a ternary gel phase. The ionic conductivity of the gels containing LiTFSI was much higher than those of the gels without LiTFSI. FT-IR study showed that cellulose triacetate interacts with Li^+ ions via their carbonyl functional groups, which may helps fast ionic conductivity. With good ionic conductivity and excellent mechanical stability, cellulose triacetate gel electrolyte is strongly recommended to be investigated further for an application in Lithium batteries.

207g Functionalized Ionic Liquids and Zwitterionic Compounds as Electrolyte Additives for Lithium Batteries

Dinh Quan Nguyen, Energy & Environment Research Division, Korea Institute of Science and Technology - KIST, 136-791 Seoul, Seongbuk-gu, Hawolgok-dong, 39-1, Seoul, South Korea, Hyun Woo Bae, Kyung Hee University, 1 Hoegidong, Dongdaemoongu, Seoul 130-701, South Korea, Hoon Sik Kim, Dept. of Chemistry, Kyung Hee University, 1 Hoegidong, Dongdaemoongu, Seoul 130-701, South Korea, Hyunjoo Lee, Energy & Environment Research Division, Korea Institute of Science & Technology, 39-1, Hawolgok-dong, Seongbuk-gu, Seoul, South Korea, and Honggon Kim, Energy & Process Research Division, Korea Institute of Science and Technology, Seoul, South Korea.

Different types of ionic liquids and zwitterionic compounds containing carbonyl-functional groups were prepared and their electrochemical properties were investigated. FT-IR study clearly demonstrates that they are able to interact with Li^+ ions through their carbonyl groups, which helps the electrochemical properties enhance. In addition, comparing with the cases of their corresponding alkyl-substituted relatives, their C-rate performance was much better upon their addition to to a non-aqueous electrolyte containing $LiPF_6$, implying the positive role of the ester functional groups.

Contaminant Dispersal in Obstructed Flow

Nicos S. Martys and Stephen J. Treado,

National Institute of Standards and Technology Gaithersburg, MD USA.

Abstract

Results are presented of a numerical study comparing the efficiency and effectiveness of contaminant removal by flushing for several rectilinear pipe flow geometries and at different Reynolds numbers.

Introduction

The quality and safety of drinking water is of paramount importance for any community. When a pipe or water conduit system is contaminated, it is important to determine the best strategies for removal of contaminants. The traditional approach for decontaminating plumbing systems is to flush with water at high volumes, primarily because this methodology is most easily implemented, and there are few simple alternatives. However, the effectiveness of this approach has not been demonstrated, and in fact, for several reasons, this may not be the best way to remove contaminants. For example, it can be argued that since high velocity water flows in piping systems are in the turbulent regime, the eddies that are generated tend to inhibit the transport and removal of entrained contaminants. This affects both the initial distribution and removal of contaminants. Further, it is not understood to what degree the location of the contaminant in the pipe system will affect its removal rate. We investigate these possibilities by comparing transport in different flow geometries and at low and high Reynolds numbers using numerical modeling. An effort is made to separate the effect of overall flow rate from the actual flow path the contaminant takes, which will change with Reynolds number. In the more complex geometries, we place "patches" of contaminant in different locations to study the ease of contaminant removal from each location.

The paper is organized as follows: first the pipe geometry is described, then a brief description of the flow model is given and the flow fields are illustrated. We then describe adding a contaminant to the system. The contaminant is allowed to undergo hydrodynamic dispersion and we track the amount remaining in the pipe over time.

Pipe geometries

To investigate the movement of contaminants, three idealized flow geometries that are representative of typical pipe systems were considered:

1) The base system for comparison is a straight square pipe.

2) The second system studied is a U shaped pipe (see Fig. 1). This system could be representative of a valve in the pipe system.
3) The third pipe system is one with a cavity or pocket (something like a trap) out of the line of the main flow (see Fig 2).

Flow model

The fluid flow was modeled using a novel computational approach based on the lattice Boltzmann method. A detailed description of this model is beyond the scope of this paper and is described in (Martys 2001). It can be shown that this model maps directly to the Navier-Stokes equations and avoids certain undesirable numerical artifacts associated with standard lattice Boltzmann methods. This approach produces velocity fields with second order accuracy in space and is easily adoptable for modeling flow in rectangular geometries - one of the main advantages of the original lattice Boltzmann method. In all cases, a no-slip boundary condition is used and the simulations were carried out in three dimensions.

The Reynolds number, (Re), is given by $\mathrm{Re} = \langle V \rangle l / \nu$ where $\langle V \rangle$ is the average velocity at the inlet, l is the channel width, and ν is the kinematic viscosity. In these simulations, Re is adjusted by varying the kinematic viscosity while fixing the flow rate. That way we can isolate the effect of the flow trajectory from the effect of simply increasing the flow velocity to increase the Reynolds number. For this paper, we will consider two Re values, 30 and 3000, which are representative of typical slow and fast flow rates in pipe systems respectively.

Figures 1 and 2 illustrate the main flow features obtained from the simulations for the U shaped pipe and cavity geometries, respectively. In both figures, the fluid enters from the left side and exits out the right side. As expected, the flow fields at low and high Reynolds number were quite different. At the lower Re (Fig.1a), the flows appeared laminar with little rotational flow. In the case of the U shaped pipe, the higher Re (Fig. 1b) flow developed significant eddies, especially around flow corners. In the case of flow past a cavity, at low Re, a single rotational flow field developed near the opening of the cavity (not shown) but at the higher Re (Fig. 2), two counter rotational flows developed. Finally, the higher Re flow typically develops a greater velocity near the surface along "straight sections" of pipe, as compared to the lower Re flow, due to the development of the usual boundary layer.

Dispersion of Contaminant

To model the dispersion of contaminants we only consider the regime where hydrodynamics is important. That is, there is no diffusion of contaminant in the fluid phase. This assumption is supported by the consideration that during flushing the flow rates are so great that diffusion of the contaminant will occur over very small lengths and hence can be taken as minimal. To track the contaminant an upwind scheme is utilized that conserves mass and has little numerical dispersion over the length and time scales considered (Martys 1994). Patches of contaminant are introduced into the system and the time evolution of their dispersion is followed. We then compare the egress time for each flow case. A simple linear coupling of contaminant with the pipe wall, where the rate of absorption is proportional to the concentration difference in the fluid and on the wall, is assumed.

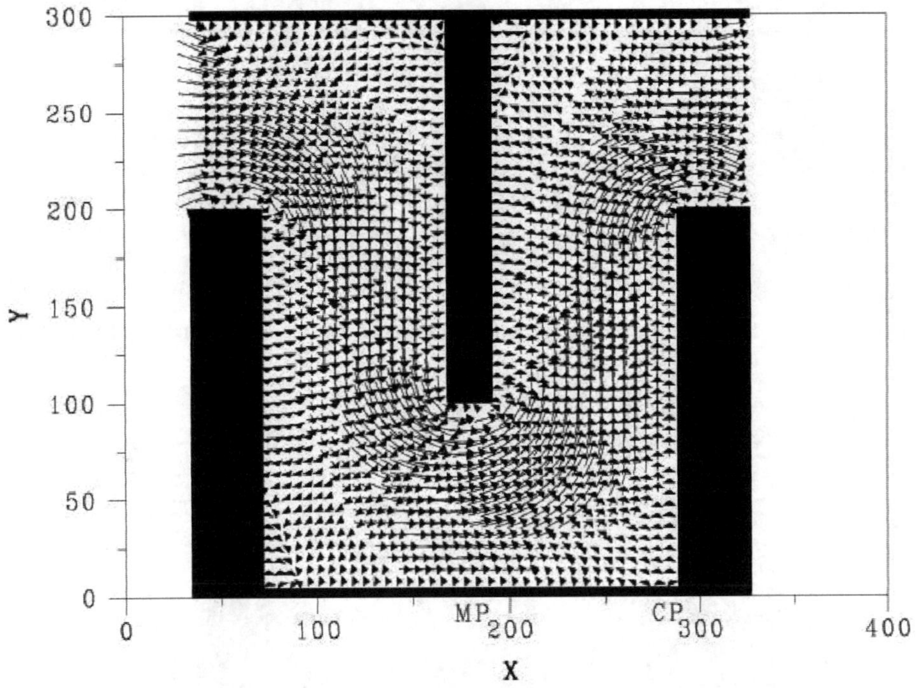

Figure (1a). The velocity fields, as indicated by the arrows, for a Reynolds number of about 30 in the U shaped pipe system. Contaminants were placed near the lower midway part of the flow path (MP) and at the bottom right hand corner (CP) of the pipe. Distance is in units of lattice spacing.

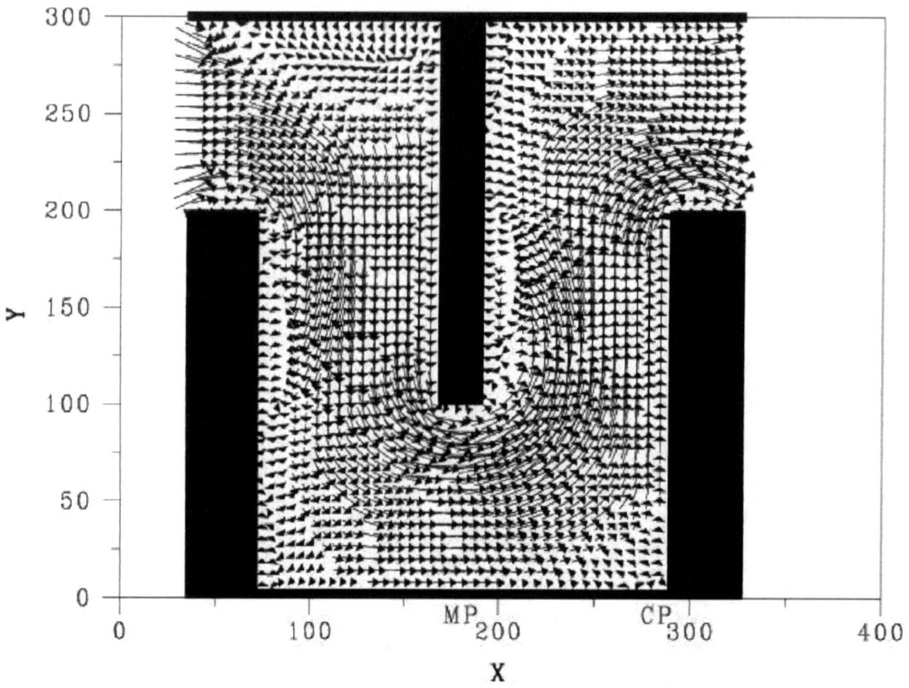

Figure (1b). Velocity fields for a Reynolds number of about 3000 for the U shaped pipe system. Note the regions of rotational flow near the corners. Patches are located near MP and CP.

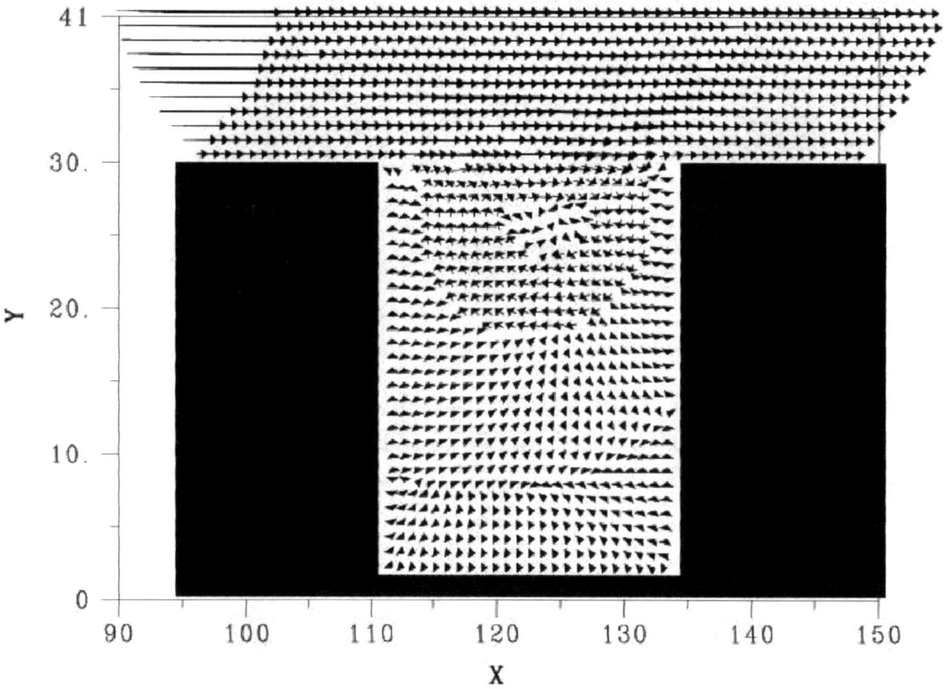

Figure 2. Flow near a cavity. Note the rotational flow near the opening of the cavity. A secondary rotation pattern developed near the bottom of the cavity. For this geometry, the contaminant was placed along the bottom of the cavity.

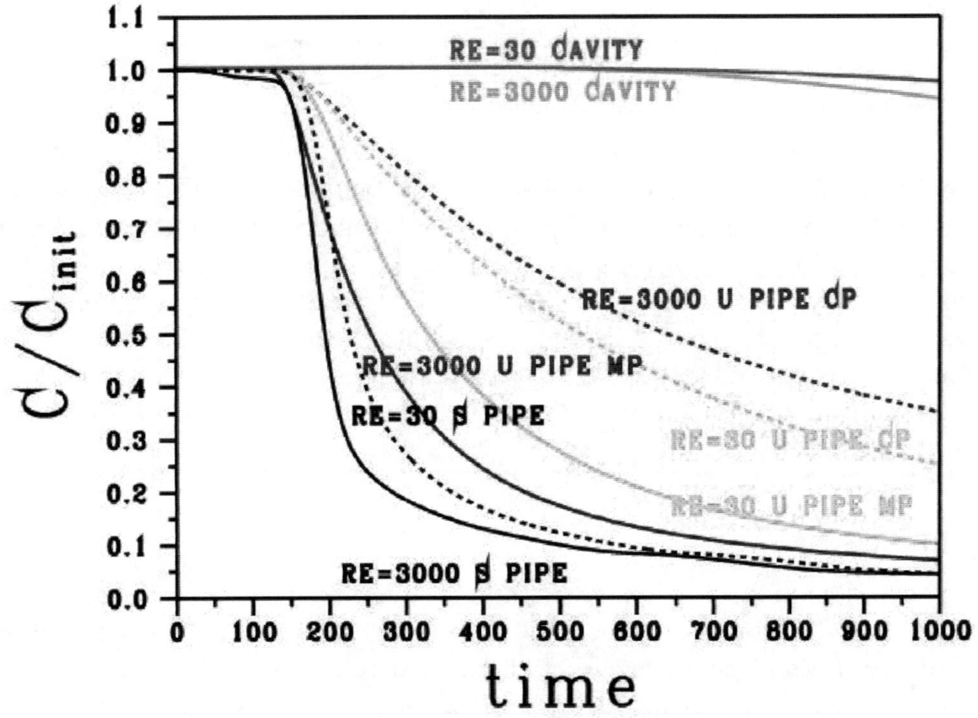

Figure 3. Normalized pipe contamination, C/C_{init} vs. time for different pipe geometries and contaminant location. Time is in given in relative units where 100 corresponds to the time it takes for the fluid to pass through the pipe system. The black lines (Re=30 dashed, Re=3000 solid) correspond to straight pipe flow (S), green and blue are for the U shaped pipe and the violet and purple for the cavity. The dashed (green or blue) lines correspond to case where the contaminant was placed in a corner (CP) and the solid lines all correspond to the contaminant placed midway (MP) in the pipe.

Figure 3 compares the amount of contaminant as a function of time for the different flow geometries and at different Re. The variable C is the total contaminant in the pipe system and is normalized by its initial value C_{init}. In the U shaped pipe systems, patches of contaminant were placed at two locations. One was near the bottom of a bend where the flow near the surface is high and a second case was considered where the contaminant was placed near the lower right corner where there was a low flow region with rotational flow (see Fig. 1). Clearly, at both Re, the contaminant left the pipe system much more efficiently when placed in the bottom straight section. When the contaminant is placed in the corner, there is a significant increase in time for its removal due to the lower flow rates. In the case of the higher Reynolds number, the delay of removal is enhanced because of the slow circulatory flow in that corner.

In the case of the flow past the cavity, the contaminant was initially placed at the bottom of the cavity. The slow rotational flow inside the cavity has a tendency to suppress the contaminant's egress. As can be seen, the concentration of contaminant remains for a significant time in the cavity. At either Reynolds number, the cavity case was the most difficult to clean. Pipe systems that contain cavities or regions out of the main flow path remain the biggest challenge as the time scale for contaminant removal was significantly higher.

While it was found that flow fields associated with higher Reynolds number can have a strong effect on the trajectory of the contaminant, it should also be noted that practice, the actual flow rates needed to achieve the higher Re will, generally, more than compensate for the reduced efficiency. For example, for the U shaped pipe, the contaminant placed near the corner took longer to egress at the higher Re. But in real applications the viscosity is fixed and the flow rate would have to increase by a factor 100 to reach the same Re value. As a result the time to reach the same level of decontamination would be reduced by a factor of 100.

Conclusion

This study has shown that location of contaminant can have a significant effect on its egress time. It was found, as a rule, that the higher the flush rate the faster the removal of contaminant; however in some cases the effect of the vortices and dead flow regions could reduce the efficiency of removal. One issue that remains is how to determine where the contaminant is most likely to reside, after initial exposure. When making estimates of needed flushing time, it is unlikely that the pipe would be uniformly contaminated. We are currently studying this issue to provide more reasonable estimates for decontamination protocols.

Acknowledgements

We gratefully acknowledge support from the U.S. EPA project manager Vince Gallardo.

References

Treado, S., Technical Issues Related to the Measurement and Modeling of Contaminant Transport and Accumulation in Building Plumbing Systems, NISTIR 7253, National Institute of Standards and Technology, Gaithersburg, MD, 20899, 8/2005.

Treado, S., Kedzierski, M., Watson, S., Bentz, D., Martys, N., Cole, K., Report of Phase One Measurement and Analysis of Building Water System Contamination and Decontamination, NISTIR 7351, National Institute of Standards and Technology, Gaithersburg, MD, 20899, 9/2006.

Martys (1994), N. S., Fractal growth in hydrodynamic dispersion through random porous media, Phys. Rev. E Vol. 50, 335-342 (1994).

Martys (2001), N. S., A Classical Kinetic Theory Approach to Lattice Boltzmann Simulation, Int. J. Mod. Phys. C, Vol. 12, No. 8, 1169-1178 (2001).

224b Yield Behavior of Wax Gel Formation in the Presence of Asphaltenes

Kyeongseok Oh, Kaushik Gandhi, Pankaj Tiwari, and Milind D. Deo. Chemical Engineering, University of Utah, 50 S. Central Campus Dr., MEB, Rm. 3290, Salt Lake City, UT 84112

High molecular weight paraffinic components have been attributed to impede the oil flow along the pipeline transport in low temperature environ by either wax deposition or wax gel formation. Wax deposition occurs even during the flow, while wax gel forms during the planned or emergency shutdown within a short period of time. It has been found that paraffinic components contribute to the evolving gel strength continuously while cooling below pour point. The contributive effect of asphaltenes to the yield stress of the gel formed was also studied. Model oils were used to determine the yield stress development in this study. Model oils were prepared by mixing mineral oil and kerosene with different amount of well-characterized waxes. The measurements of WAT and PP were performed using FT-IR and ASTM methods. As the WAT and PP are dependent on wax amount and wax quality, this study examined the WAT and PP by comparing model oils with different wax composition as well as different wax amount. Vane method was adapted to measure the yield stress at different temperatures. Constant cooling rate and consistent holding time were employed for the yield stress. The yield stress was recorded with decreasing temperatures. It was shown that the yield stress is strongly dependent on the wax amount and wax composition. Increase in yield stress values with decreasing temperature was greater for higher amount of wax in model oil indicating the gel strength is dependent upon the wax amount in the model oil. The x-intercept values obtained from yield stress versus temperature were interpreted as no-flow point, which may use as pour point alternatives. The relation between wax amount and yield stress was analyzed using FT-IR. Asphaltene used in this study is from Rangely field in northwestern Colorado. Examination of asphaltene addition was performed with 0.01 wt. % and 0.1 wt. % in model oil. Model oils were prepared with mineral oil, waxes, toluene, and asphaltene for this effort. Yield stress increases linearly decrease in temperature in asphaltene-free model oil. Asphaltene additions result in pour point reductions, 1 C in 0.01 wt%, and 4 C in 0.1 wt%. Small amount of asphaltene (0.01 wt. %) also played a significant role in yield stress reduction. A considerable reduction in yield stress was observed by asphaltene addition. A departure form the linear trend in increase of yield stress with decreasing temperature was observed at lower temperatures in the model oils with asphaltenes.

224c Coupled Combustion and Thermal Stress Computation

Naseem Ansari and Jaydeep Kulkarni. Technical Services, ANSYS Inc., Canonsburg, PA 15137

Strict air pollution regulations in recent decades have been a driver for advanced low NOx burner development. The availability of computational fluid dynamics tools have accelerated this effort by enabling rapid investigation of many virtual designs in order to optimize the temperature and pollutant production field before going to physical prototype and test furnaces.

In addition to the challenges of pollution control and optimal heat transfer for steam production or process heat in combustion energy systems, creep and fatigue due to thermal stresses are another major concern of design engineers. Historically, information about flame characteristics or surface temperatures have been passed from combustion engineers to stress engineers, who independently applied those conditions in a structural analysis. Advanced tools that combine these two disciplines are becoming more commonplace.

In this study, coupled computational fluid dynamics and finite element analysis are used to study flame shape, NOx production, and nozzle thermal stress in a coal-fired low NOx burner. Methodology, solution time, and best practices are covered.

Mass Transfer with coaxially placed entry region coil – disc assembly as turbulence promoter in vertical tube flow reactor

Vaka Murali Mohan., Rajendra Prasad P*., Sujatha, V and Sarveswara Rao, S

Electro Chemical Engineering Research laboratory,

Department of Chemical Engineering, Andhra University, Visakhapatnam-03, A P, INDIA.

Abstract

Studies on mass transfer with coaxially placed entry region coil - disc assembly as turbulence promoter in vertical tube flow reactor have been studied in this paper. The electrolyte was equimolal Potassium ferricyanide, Potassium ferrocyanide and excess sodium hydroxide. The mass transfer correlation is based on law of the wall similarity. The variables covered in this study are the flow rate of the electrolyte, the geometric parameters of the promoter – pitch of the coil, length of the coil, disc diameter and height of the disc and velocity of the fluid. Mass transfer coefficient is increasing with velocity and with disc diameter. Mass transfer coefficient is increases up to a maximum there after it shows a decrease with increase in height of the disc. Pitch of the coil increases to maximum and then showed a decreasing trend. Mass transfer coefficient increases with increase in length of the coil because of induced swirl motion of the fluid. The increase continued to a maximum value at coil length of 0.095 m, beyond which a marginal decrease was observed. Augmentations achieved are range from 1.5 to 8 fold as the velocity varies from 0.0591 m/s to 0.2751 m/s. A model was developed in terms of geometric parameters and presented.

$$\bar{g} = 5.76\ [\mathrm{Re}_m^+]^{-0.003} \left(\frac{P_c}{d}\right)^{0.003} \left(\frac{L_c}{d}\right)^{0.001} \left(\frac{d_d}{d}\right)^{0.616} \left(\frac{H_d}{d}\right)^{-0.00001}$$

Keywords: entry region coil, disc, turbulence promoter, mass transfer, modeling.

1. Introduction

Flow of electrolyte through circular conduit with axially placed insert promoter generates a variety of flow fields. The turbulence promoters have good potential in electro chemical reactors namely electro winning, refining, electro organic synthesis etc. Lin, C, S et al [1] studied mass transfer rates of ions and other reacting species in diffusion controlled reactions in order to obtain design information for electrochemical processes. The flow devices form an important group of passive augmentation methods which coaxially placed entry region coil - disc assembly as turbulent promoter. In pursuit of this, several strategies have been devised and

* Prof. Dr. P.Rajendra Prasad
Department of Chemical Engineering,
A U College of Engineering, Visakhapatnam - 03, A P, INDIA
E-mail : murali_vaka@yahoo.com, Tel. Ph no. 91-891- 2844899. Fax no: 91-891-2747969.

adopted. Chiou, J, P [2] made experimental investigation for the augmentation of heat transfer coefficient in forced convection heat transfer. He employed spiral spring coil as an insert promoter in circular tube. Changal Raju, D [3] obtained data for mass transfer at wall electrodes using limiting current technique. Coils of different sizes wound on central annular rod were used as promoter. Paisarn Naphon [4] investigated the heat transfer characteristics and the pressure drop of the horizontal double pipe with coil-wire insert. Alberto Garcı´a et al [5] studied wire coil inserts offer their best performance within the transition region. Rajendra Prasad, P [6, 7] studied ionic mass and momentum transfer with coaxially placed spiral coils as turbulent promoter in homogeneous flow and in fluidized beds and did experiments using ferri - ferro cyanide system by limiting current technique. Discs placed across the flow of circular conduits generate wakes and internal circulation within the column. The effectiveness of a disc varies with the length of the column. Venkateswarlu, P and Jagannadha Raju, G, J [8, 9] obtained mass and momentum transfer data with co-axially placed discs on a central rod. The augmentation was about 12 fold over the data in conduits in the absence of disc promoter elements. Ravi et al [10] obtained mass transfer data on the walls of the fluidized beds using an electrochemical technique with disc as promoter. Sherwood, J, D and Stone, H, A [11] determine the flow of fluid around a disc in a pipe and the added mass of the accelerating disc and flow profiles are also presented. Ziókowska, I and Dolata, M [12] investigated the intensification of the heat transfer between a gas stream and the wall of a tube provided with a set of perforated discs used as turbulence promoters and the accompanying pressure drop. Mass transfer with entry region coil-disc assembly as turbulence promoter was selected for the present study.

2. Experimentation

The experimental unit essentially consisted of a re-circulating tank, a centrifugal pump, rotameter and an entrance calming section, a test section, and an exit calming section. The test section was made with Perspex material and was provided with a number of point electrodes fixed flush with its surface longitudinally at equal spacing (0.03 m). The wall of the entrance calming section acted as wall electrode in much larger surface area. The promoter was mounted in the test section coaxially by means of gland nuts and positioned by supporting grid. Entry region coil – disc assembly as turbulence promoter was made of copper wire with provision to fix it rigidly within test section. The details of the coil-disc promoter were shown in Fig. 1. After inserting the promoter in the column, one-hundred liter of the electrolyte consisting of equimolal 0.01M potassium ferricyanide and potassium ferrocyanide together with 0.5N sodium hydroxide was prepared in the storage tank. The concentration of the electrolyte is maintained constant throughout the experiment, small changes of temperature of electrolyte offers changes in Schmidt number in the range from 902.5 – 1089. The solution is kept in the dark and closed recirculation tank. The electrolyte was pumped through the test section. The flow rate of the electrolyte was adjusting by operating the control and by pass valves. Limiting current data were measured at point copper electrodes for the reduction of potassium ferricyanide ion. The procedure followed similar to the works of Lin et al [1] and several others [3, 6 and 7]. Room temperature is around 30^{0}C. Temperature is recorded for every reading of limiting current using the thermometer with 0.1^{0}C accuracy. This ensures the accuracy of physical properties such as μ, D, ρ. Thermometers are attached at both ends of the column. The electrochemical reaction involved is given below:

$$Fe\,[(CN)_6]^{-3} + e^- \rightarrow Fe[(CN)_6]^{-4} \qquad \text{(a)}$$

The limiting current was indicated by a small increase in current for a sharp increase in voltage. The experiments were repeated by varying the pitch of the spiral coil (P_c), Length of the coil (L_c), Diameter of the disc (d_d) and Height of the disc (H_d). The range of variables covered is given in Table 1. Mass transfer coefficients are calculated by the following expression:

$$k_L = \frac{I_L}{n\,A\,F\,c_i} \qquad (b)$$

3. Theory

Flow of electrolyte through tubular flow reactor with entry region coil insert promoters will establish momentum and concentration profiles and was shown in Fig 2. The profile is divided into two regions, the inner region and outer region. The inner region again divided into laminar sub layer and viscous buffer zone. The outer region is turbulent core region and is fully turbulent. The turbulent core separated by an interface. A model is developed with the following assumptions:

1. The fluid is Newtonian, incompressible, and has constant transport properties and independent of time.

2. Flow through smooth tube is fully developed with no entrance effects. The effect of body forces is small in comparison to that of viscous and inertial forces.

3. The axial velocity profile is independent of axial coordinate, similarly the concentration gradients.

4. The mass flux varies in the radial direction. The mass flux varies linearly with y, and is represented by

$$\frac{N}{N_w} = 1 - \frac{y}{R} \qquad 1$$

The radial variation of mass flux is taken into account by dividing into viscous-buffer zone and turbulent core. The flux at the interface is considered radially opposite directions, Flux at the interface towards the wall,

$$N_{i(v-b)} = k_{(v-b)} (c_i - 0) \qquad 2$$

Flux at the interface towards the axis,

$$N_{i(t)} = -k_{(t)} (c_i - c_b) \qquad 3$$

5. The shear stress is constant in the radial direction. Deissler, R, G and Eian, C, S [13] have found that the velocity and concentration profiles are relatively insensitive to shear stress distribution for fully developed turbulent fluid flow, hence can be assumed constant.

6. In the viscous buffer zone, the eddy diffusivity and eddy viscosity ratio varies along with the distance from the wall and is independent of Sc. Direct numerical simulations presented by Lyons et al. [14] suggest Sc to be unity at the wall. Hence, in the present case it is assumed that as $y^+ \rightarrow 0$, Sc=1.

7. The eddy diffusivity controls transfer process in viscous-buffer zone, while eddy viscosity controls the transport in turbulent core.

4. Model Development

In turbulent core, the intensity of eddies will be higher than the v-b zone. Therefore, during the transfer process, the resistance to the exchange of mass between the bulk fluid and the solid surface is often confined to the viscous and buffer region. The total resistance to the transfer process is the sum of the resistances in the viscous-buffer zone and turbulent layer proposed by Aravinth, S [15] and is represented by,

$$\frac{1}{k} = \frac{1}{k_{v-b}} + \frac{1}{k_t} \qquad 4$$

Equation 4 represents a two-layer resistance separated by an interface, similar to that proposed by Hugh mark, [16]. The total resistance in the v-b zone and the turbulent core is represented by equation 4 and are similar to the boundary layer resistance and the resistance corresponding to random eddies in the region beyond the boundary layer. For a fully developed turbulent flow, the shear stress and the heat flux can be written as,

$$\tau = (\mu + \rho\varepsilon)\frac{du}{dy} = \rho(\nu + \varepsilon)\frac{du}{dy} \qquad 5$$

$$N = (D + D_e)\frac{dc}{dy} \qquad 6$$

In the viscous-buffer region,

Equation 6 on integration with the boundary conditions: at y=0 , c = 0 and at $y = y_i^+$, $c = c_i$ and substitute equation 5 in 6 gives.

$$\frac{\Delta c}{N_{i(v-b)}} = \frac{1}{\sqrt{\tau_0/\rho}} \int_0^{y_1^+} \frac{dy^+}{\frac{1}{Sc} + \frac{D_e(y^+)}{\upsilon}} \qquad 7$$

Combining equation 2 and 7 yields,

$$\frac{1}{k_{v-b}} = \frac{1}{\sqrt{\tau_0/\rho}} \int_0^{y_1^+} \frac{dy^+}{\frac{1}{Sc} + \frac{D_e(y^+)}{\upsilon}} \qquad 8$$

In the turbulent region,

As the transport is mainly by eddy means, the molecular diffusivity and the viscosity can be neglected in the turbulent core. Combining equation 5 and 6 and integrating along the turbulent layer with the boundary conditions: $c=c_b$ at $u = u_b$ and $c=c_i$ at $u = u_i$, yields,

$$\frac{dc}{du} = \frac{N_{i(t)}}{\tau_0/\rho}\left(\frac{\nu_e}{D_e}\right)_t \qquad 9$$

Substitute equation 3 in 9.

$$\frac{1}{k_t} = \int_0^{y_1} \frac{-du}{\tau_0/\rho}\left(\frac{\nu_e}{D_e}\right)_t = \frac{-(u_i - u_b)}{\tau_0/\rho}.\Phi \qquad 10$$

$$\frac{1}{k_t} = \frac{\Phi u_b}{\tau_0/\rho}\left(1 - \frac{u_i/u^*}{u_b/u^*}\right) = \frac{\Phi u_b}{\tau_0/\rho}\left(1 - \frac{u_i^+}{u_b^+}\right) \qquad 11$$

In turbulent core $u_m \cong u_b$, $\qquad u_m^+ = u_b^+ = \sqrt{2/f} \qquad 12$

Substitute equation 12 in 11

$$\frac{1}{k_t} = \frac{\Phi u_b}{\tau_0/\rho}\left(1 - u_i^+\sqrt{f/2}\right) \qquad 13$$

Equation 8 and 13 can be added using the resistance in series definition, i. e equation 4

$$\frac{1}{k} = \frac{1}{\sqrt{\tau_0/\rho}} \int_0^{y_1^+} \frac{dy^+}{\frac{1}{Sc} + \frac{D_e(y^+)}{\upsilon}} + \frac{\Phi u_b}{\tau_0/\rho}\left(1 - u_i^+\sqrt{f/2}\right) \qquad 14$$

Applying the assumption 7 that eddy viscosity controls the transfer process in turbulent core and the ratio $\Phi = 1$.

$$\frac{1}{k} = \int_0^{y_1^+} \frac{dy^+}{\frac{1}{Sc} + \frac{D_e(y^+)}{\upsilon}} + \sqrt{2/f} - u_i^+ \qquad 15$$

$$\frac{1}{k} = \frac{1}{St.\,u_b} = \frac{1}{St.\sqrt{2/f}} = \frac{\sqrt{f/2}}{St} \qquad 16$$

Substitute equation 16 in 15.

$$\frac{f/2St - 1}{\sqrt{f/2}} + u_i^+ = \int_0^{y_1^+} \frac{dy^+}{\frac{1}{Sc} + \frac{D_e(y^+)}{\upsilon}} = g(\mathrm{Re}_m^+, Sc) \qquad 17$$

The function g is written as product of $\bar{g}$ and F (Sc).

$$g(\mathrm{Re}_m^+, Sc) = \bar{g}\,(\mathrm{Re}_m^+)\,F(Sc) \qquad 18$$

The law of wall can also represent well for the present study also. The integral of equation 16 is only dependent on Sc only and is represented by F (Sc). The F(Sc) expression obtained in Prandtl and Von Karman analogy is chosen for the present study. It is also supported by Deissler, R, G [17] and Sparrow, E, M and Malina, J, A [18] for Sc > 1 the empirical equation of F (Sc) is

$$F(Sc) = 9.3(Sc - 1)Sc^{-1/4} \qquad 19$$

$$\mathrm{Re}_m^+ = \frac{d_d}{d}\,.\,\mathrm{Re}.\,\sqrt{\frac{f}{2}} \qquad 20$$

The function $\bar{g}$ is influenced by Re_m^+ and geometric parameters.

$$\bar{g}\,(\mathrm{Re}_m^+, \Phi_1, \Phi_2, \Phi_3)$$

The resultant function can be represented as

$$\bar{g} = C\,\mathrm{Re}_m^{+\,n_1}, \Phi_1^{\,n_2}, \Phi_2^{\,n_3}, \Phi_3^{\,n_3}$$

$\Phi_1 = P_c/d$, $\Phi_2 = L_c/d$, $\Phi_3 = d_d/d$ and $\Phi_4 = H_d/d$

Data yielded the following equation for entry region coil -disc assembly as turbulence promoter.

$$\bar{g} = 5.76\ [\mathrm{Re}_m^+]^{-0.003} \left(\frac{P_c}{d}\right)^{0.003} \left(\frac{L_c}{d}\right)^{0.001} \left(\frac{d_d}{d}\right)^{0.616} \left(\frac{H_d}{d}\right)^{-0.00001} \qquad 21$$

Standard deviation = 1.92, Average deviation = 1.83.

Equation 21 is compared with the data of Rajendra Prasad, P [6, 7] and Venkateswarlu, P [8, 9] and shown in Fig.3. The figure reveals that the data of Rajendra Prasad, P and Venkateswarlu, P is well fitted in this correlation.

5. Results

Entry region coil-disc assembly as turbulence promoter gives considerable energy losses because of the circulation generated by the coil as well as disc generated a special flow pattern which is particularly suitable for electro processes like electro organic synthesis, electro winning, electro refining, electro dialysis, reverse osmosis and several other processes including bio reactors. Local mass transfer coefficients were computed with measured limiting currents along the column and presented as Fig.4. The mass transfer coefficients were decreasing along the length of the conduit. The figure reveal, as the velocity increases mass transfer coefficients were also increased. The figure also contains the data of Lin et al which is a smooth pipe flow. The observations reveal mass transfer coefficients were decreasing with distance up to a value of 0.15 m beyond which mass transfer coefficient values reaches a steady state value and coincide with Lin et al.

Integral average mass transfer coefficient values are computed and plotted against distance and shown as Fig.5. Integral average mass transfer coefficients values are decreasing with distance along the length of the conduit. The figure also contains the data of Lin et al which is a smooth pipe flow. The figure reveal, as the velocity increases mass transfer coefficients were also increased. The figure reveals that the mass transfer coefficients were decreasing with distance up to a value of 0.15 m beyond which mass transfer coefficient values reaches a steady state value and coincide with Lin et al.

Geometric parameters has strong influence on mass transfer, mass transfer coefficient versus velocity data with best augmenting parameters is drawn as Fig. 6. The figure reveal, as the velocity of the fluid increases, mass transfer coefficient increases. Higher augmentations achieved are range from 2 to 10 folds as the velocity varies from 0.0591 m/s to 0.2751 m/s.

6. Conclusions

The mass transfer coefficients were decreasing with distance along the length of the conduit. The maximum mass transfer coefficient was observed at 0.15 m length of the conduit with in the range of parameters of the promoter covered in the study. As the velocity of the fluid increases, mass transfer coefficient increases. Higher augmentations achieved are range from 2 to 10 folds as the velocity varies from 0.0591 m/s to 0.2751 m/s. A model was developed in terms of geometric parameters and presented.

$$\bar{g} = 5.76\ [\mathrm{Re}_m^+]^{-0.003} \left(\frac{P_c}{d}\right)^{0.003} \left(\frac{L_c}{d}\right)^{0.001} \left(\frac{d_d}{d}\right)^{0.616} \left(\frac{H_d}{d}\right)^{-0.00001}$$

7. List of symbols

A = Area of electrode, m^2
c = concentration of electrolyte, kg-mole/ m^3
c_l = length of the conduit, m
d = diameter of the conduit, m
d_d = diameter of the disc, m
D = diffusion coefficient, m^2/s

D_e = eddy diffusivity, m^2/s
F = faraday's constant = 96,500 coulombs/ g-mole
H_d = height of the disc, m
I_L = limiting current density, amps
k_L = mass transfer coefficient, m/s
L_c = Length of the coil, m
N = mass flux
P_c = Pitch of the coil, m/turn
R = radius of the conduit, m
u_i = velocity at the interface, m/s
u_b = average velocity, m/s
u^* = friction velocity, $\sqrt{\tau_0/\rho}$
u_i^+ = dimensionless velocity, $u_i^+ = 2.5\log(y/e) + R(h^+)$
$R(h^+) = 2.5\log(2e/D) + \sqrt{2/f} + 3.75$
u_b^+ = dimensionless bulk velocity
u_m^+ = average fluid velocity, m/s,
V = superficial velocity, m/s
y = coordinate distance normal to wall, m
y_1 = distance from wall at which $u = u_b$
y_1^+ = dimensionless distance, yu^*/v

Dimensionless

$\bar{g}$ = modified mass transfer function
Re_m^+ = modified Reynolds number
St = Stanton number, k/u_b or k_L / v
Sc = Schmidt number, ν/D
Φ = Ratio between molecular viscosity and eddy viscosity, $(\nu_e/D_e)_t$
n =number of ions transferred

Greek symbols

τ = shear stress, kg/ms^2
μ = viscosity of the fluid, poise
ρ = density of the fluid, kg/m^3
τ_0 = wall shear stress, kg/m s^2, $f/2.\ \rho.u_b^2$
υ = kinematics viscosity
υ_e = eddy kinematics viscosity

Subscripts

b = buffer
i = interface
t = turbulent
o = wall
v = viscuous
v-b = viscuous buffer region

Appendices References

[1]. Lin, C, S., Denton, E, B., Gaskill, H, S and Putnam, G, L "Diffusion controlled electrode reactions" *Industrial and Engineering Chemistry, vol 43/9, Sep 1951, pp 2136-2143.*

[2]. J.P.Chiou, "Experimental investigation of the augmentation of forced convection heat transfer in a circular tube using spiral spring inserts" *Trans.A.S.M.E., vol 109, (1987) 300.*
[3]. D. Changal Raju, "Augmentation of mass transfer at the outer wall of concentric annuli in absence and presence of fluidizing solids effect of wires wound on central rod". *Ph.D. Thesis, Andhra University, Visakhapatnam, 1984.*
[4]. P. Naphon, "Effect of coil wire insert on heat transfer enhancement and pressure drop of the horizontal concentric tubes" *Int. Comm. of Heat and Mass Transfer, vol. 33, Iss 6, (2006) 753.*
[5]. A. Garcı´a, P. G. Vicente and A. Viedma, "Experimental study of heat transfer enhancement with wire coil inserts in laminar transition turbulent regimes at different Prandtl numbers" *Int. J. Heat and Mass Transfer, vol. 48, (2005) 4640.*
[6] P. Rajendra Prasad, "Studies on ionic mass transfer with coaxially placed spiral coils as turbulence promoter in homogenous flow and in fluidised beds" *Ph.D Thesis, Andhra university, INDIA, 1993.*
[7]. P. Rajendra Prasad., V.Sujatha., C, B, Sarma and G J V J Raju, "Studies on ionic mass transfer in the presence of spiral coil turbulence promoter in batch fluidized beds", Chemical Engineering and Processing, Vol.43, (2004) 1055.
[8] P. Ventateswarulu and G. J. V. J. Raju, "Studies on ionic mass transfer with coaxially placed discs as turbulence Promoter" *Indian Chemical Engineer Journal, vol 31, (1990) 101.*
[9] P. Venkateswarlu, "Studies on ionic MT with coaxially placed discs on a rod as turbulence Promoter" *Ph.D., Thesis, Andhra University, Visakhapatnam, INDIA, 1987.*
[10] T. Ravi., B. Srinivasa Rao., P. Gopala Krishna and P. Venkateswarlu, "Ionic mass transfer studies in fluidized beds with coaxially placed discs on a rod as internal" *Chemical Engineering and Processing Journal, vol 35, (1996) 187.*
[11] J. D. Sherwood and H. A. Stone, "Added mass of a disc accelerating within a pipe" *Physics of Fluids Journal, vol 9, (1997) 3141.*
[12] I. Ziókowska and M. Dolata, "Heat and momentum transfer in gas flowing through heated tube with turbulence promoters", *Int. J of Heat and Mass Transfer, vol 37,(1994) 1839.*
[13] R.G. Deissler, C.S. Eian, "Analytical and experimental investigation of adiabatic turbulent flow in smooth tubes" , *Nat. Adv. Comm. Aeronaut., Washington, DC, TN-2629, 1952.*
[14] S.L. Lyons, T.J. Hanratty, J.B. McLaughlin, "Direct numerical simulations of passive heat transfer in turbulen channel flow", *Int. J. Heat and Mass Transfer, vol 34 (1991) 1149.*
[15] S.Aravinth, "Prediction of heat and mass transfer for fully developed turbulent flow through tubes" *Int. J of Heat and Mass Transfer Vol 43 (2000) 1399.*
[16] G.A. Hughmark, "Momentum, heat, and mass transfer analogy for turbulent flow in circular pipes", *Ind. Eng. Chem. Fundamentals, vol 8 (1969) 31.*
[17] R.G. Deissler, "Analysis of turbulent heat/ mass transfer and friction in smooth tubes at high Prandtl & Schmidt". *National Adv. Comm. Aeronaut, Washington, DC, TN (1955) 1210.*
[18] E.M. Sparrow, J.A. Malina, "Variable property, constant-property, and entrance-region HT results for turbulent flow of water and oil in a circular tube", *Chemical Engineering Science Journal, vol 19 (1964) 953.*
[19] D.F.Dipprey and R.H.Sabersky, "Heat and momentum transfer in smooth and rough tubes at various Prandtl numbers" *Int. J of Heat and Mass Transfer, vol 6 (1963) 329.*

Appendices Table.1 Range of variables covered in the present study

Variable	Minimum	Maximum
Pitch of the coil, m/turn	0.015	0.035
Length of the coil, m	0.035	0.125
Disc diameter, m	0.02	0.04
Height of the disc, m	0	0.50
Velocity, m/s	0.0591	0.2751

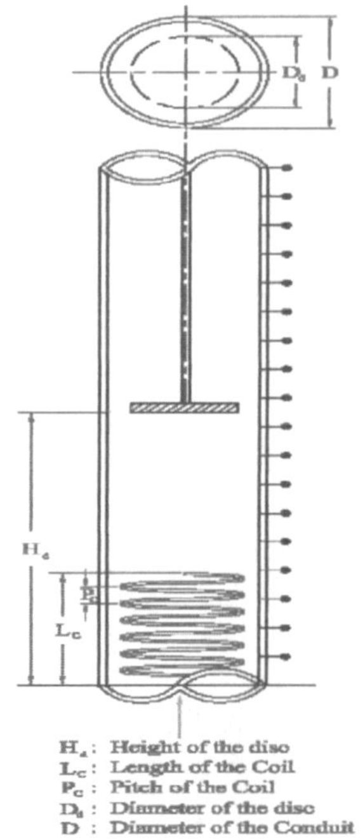

Fig.1 Details of the test section

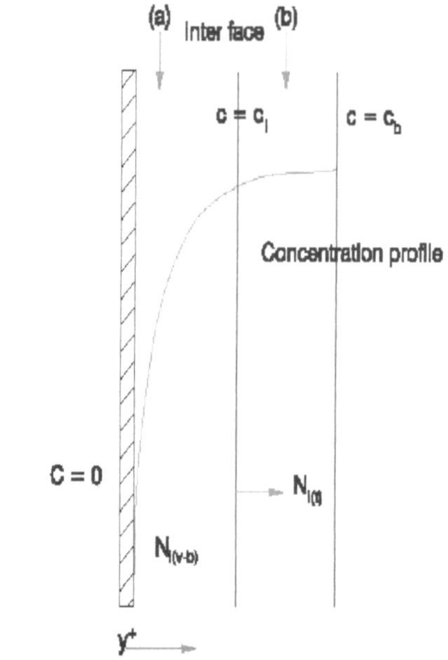

Fig.2 Concentration changes in(a) viscous buffer zone (b) turbulent core

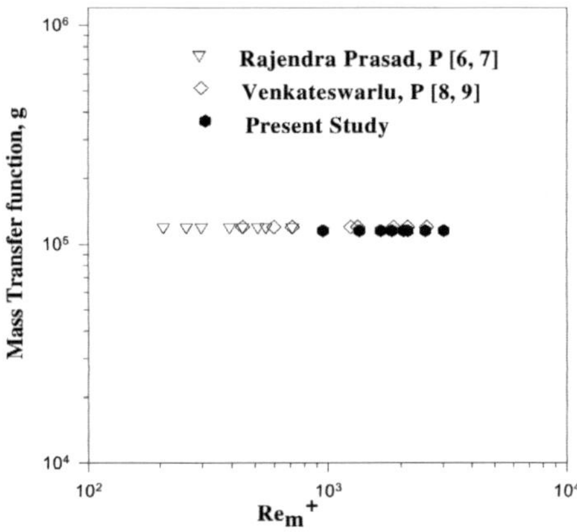

Fig.3 Mass transfer correlation comparission graph

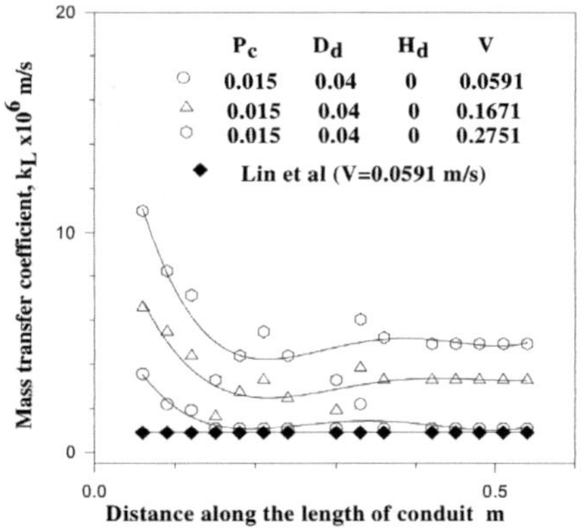

Fig.4 Variation of k_L with along the length of the conduit

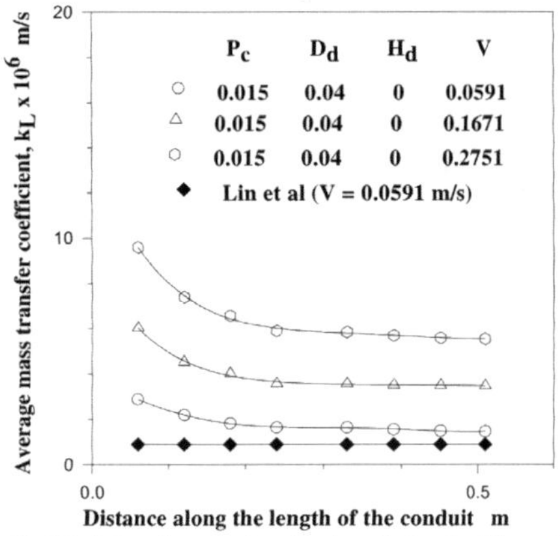

Fig.5 Variation of average k_L with along the length of the conduit

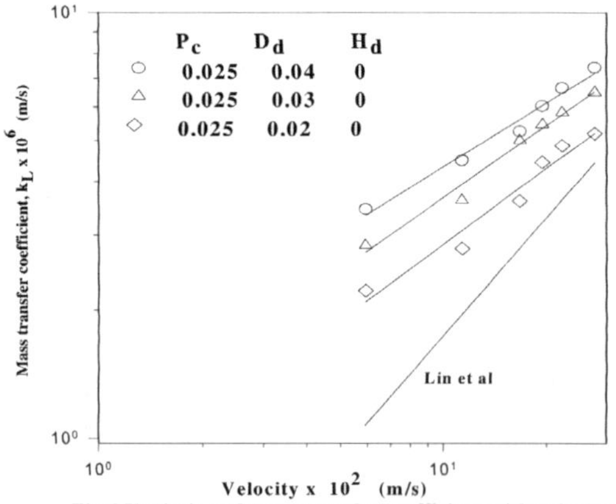

Fig.6 Variation of mass transfer coefficient with velocity

Damped Wave Conduction and Relaxation in High Temperature Thermal Barrier Coatings

Dr. Kal Renganathan Sharma[1] PE

Prairie View A & M University

Prairie View, TX 77446

Nomenclature

$c_1, c_2, ..c_n$	*integration constants*
g	*space domain portion of the solution for wave temperature in naval warhead*
h	*time domain portion of the solution for wave temperature in coating*
$J_0(x)$	*Bessel function of the first kind and zeroth order*
k_c	*thermal conducitivity of coating* (w/m/K)
k_n	*thermal conductivity of naval warhead* (w/m/K)
T_c	*temperature in coating* (K)
T_n	*temperature in naval warhead* (K)
r	*radial direction* (m)
R	*radius of the naval warhead* (m)
t	*time* (s)
u_c	*dimensionless temperature in coating* $(T_c - T_{fire})/(T_0 - T_{fire})$
u_n	*dimensionless temperature in naval warhead* $(T_n - T_{fire})/(T_0 - T_{fire})$
$V(\tau)$	*time domoin of the solution for wave temperature in naval warhead*
V_{ch}	*velocity of heat in coating* (m/s)
V_{nh}	*velocity of heat in naval warhead* (m/s)
w_c	*dimensionless wave temperature in coating*
w_n	*dimensionless wave temperature in naval warhead*
X	*dimensionless distance* $(r/(\alpha_{nr}\tau_r)^{1/2})$
X_R	*dimensionless radius* $(R/(\alpha_{nr}\tau_r)^{1/2})$
$X_{R+\delta}$	*dimensionless outer radius of naval warhead and coating assembly* $(R+\delta)/(\alpha_{nr}\tau_r)^{1/2}$
$Y_0(x)$	*Bessel function of the second kind and zeroth order*

[1] Dr. Kal Renganathan Sharma PE, Adjunct Professor, Department of Chemical Engineering, PO Box 519, MS 2505, Room 201A Wilson Building, Prairie View A & M University, Prairie View, TX, 77446. Tel: (936) 261 9413; Fax: (936) 261 9419; Email: jyoti_kalpika@yahoo.com

Greek

α_{cr}	*thermal diffusivity in naval warhead* (m^2/s)
α_{nr}	*thermal diffusivity in naval warhead* (m^2/s)
β	*ratio of relaxation times* (τ_{cr}/τ_{nr})
δ	*thickness of coating* (m)
γ	*ratio of thermal diffusivities* (α_c/α_n)
λ_n	*Eigen values*
ϕ	*space domain portion of the solution in the wave temperature in coating*
τ_{cr}	*relaxation time of naval warhead (s)*
τ_{nr}	*relaxation time of naval warhead* (s)

I. INTRODUCTION

Thermal barrier coatings, TBC, are advanced material systems applied to metallic surfaces operated at elevated temperatures. These coatings serve as insulators of heat. TBC can be used in naval warheads. In the event of fire, metallic bulkheads and decks spread fire by heat transfer which increases the temperature on the back side above the ignition temperature of common combustibles in adjacent and overhead compartments. Inorder to keep temperatures on the back side below the ignition temperature of the common combustibles, fire insulation such as mineral wool ot Structogard is used by the Navy. The role of the TBC is to prevent firespread. Coated onto a combustible substrate they can prevent the ignition and spread of flame. These can be used in applications with high fire potential. Magnesium oxychloride sorrel cement coatings have been developed by Delphi Research[1] to be used as a barrier system. The ceramic coating can be prepared using chemical vapor deposition process or spray coating process. The thermal conductivity of the coating may be reduced by increasing the porosity of the coat. Magnesium oxychloride cement coating co-bonded with high alumina calcium aluminate cement colloidal silica was found to be effective. One of the critical parameters in the design of such systems is the resistance offered by the material to temperature rise fires. ASTM standards discuss this parameter in the UL 1709 test[2]. UL 1709 test lays down the performance criteria for insulation material. The backside average temperature rise should not exceed 250 F above the ambient when tested under the conditions prescribed in the test. UL-1709 fire is specified at 2000F and 200

kw/m^2 for 30 minutes or more. The thrust of the research in development of TBC is looking to advance fire insulation technologies and bring in new materials that reduce weight and cost.

The nature of heat transfer during the firespread is transient. The insulation materials are usually heterogeneous in nature. Under short timescales the Fourier's law of heat conduction does not accurately depict the real events. Mathematical models that can predict the time of rise of insulation temperature specified in UL 1709, that use Fourier parabolic law of heat conduction cannot be relied upon. This is because of the presence of "blow-up" or singularities in the expression for heat flux in the analytical solution obtained using Fourier's law. Damped wave conduction and relaxation may become significant during the short time scales expected when TBC is used to delay onset of fire. It cannot be neglected.

Generalized Fourier's law of heat conduction can be used to develop analytical solutions that do not have any singularities at short time scales[3]. Six reasons were given in Sharma[4], to seek a generalized Fourier's law of heat conduction. Sharma [5-8] discussed the manifestation of the damped wave transport and relaxation equation in industrial applications and provided bounded solutions within the constraints of second law of thermodynamics. It was shown that the generalized Fourier's law of heat conduction can be derived by including the acceleration term in the free electron theory, the acceleration term in the Stokes-Einstein theory for molecular diffusion, by accounting for the accumulation term in the kinetic theory of gases and combining in series the Hooke's elastic element and Newton's viscous element in the viscoelastic theory. The relaxation time was found to be a third of the collision time of the electron and the obstacle. The velocity of heat was found to be identical with the velocity of mass derived from kinetic representation of pressure or the Maxwell representation of the speed of molecules.

Reference to the generalized Fourier's law of heat conduction can be traced back to Maxwell [9], Morse and Feshbach [10]. Cattaneo [11] and Vernotte [12] postulated this equation independently. This equation can be used to account for finite speed of heat . Reviews in the literature of the use of generalized Fourier's law of heat conduction have been provided by Joseph and Preziosi [13] and Ozisik and Tzou[14]. Micro to macro scale behavior has been discussed by Tzou[15]. Early investigators who tried to measure the relaxation time parameter reported values that were low of the order of a few nanoseconds. In applications, such as in a firespread, for practical purposes this effect can be neglected should this be the case. However recent experimental evidence presented by Mitra and Kumar et. al.[16] for materials with non-homogeneous inner structure show that the relaxation time can be of the order of 16-20 seconds. This means that the finite speed of heat conduction effects have to be estimated in the insulation systems for warheads.

Some early investigators have raised some concerns about violations of second law of thermodynamics by the generalized Fourier's law of heat conduction (Bai and Lavine[17], Taitel[18], Barletta and Zanchini[19], Zanchini[20]). Taitel[18] attempted to obtain analytical solution to the governing equation and found that the solution temperature for some values exceeded the boundary temperature indicating a possible violation of Clausius inequality. Sharma[4] presented an analytical solution for the case of a finite slab subject to constant wall temperature. The final condition in time as the fourth condition for the second order hyperbolic PDE governing equation was shown to result in well bounded solutions. This means that care must go into the choice of the conditions used in the boundaries of space and initial and final time values. They have to be physically realistic such as at steady state equilibrium temperature is attained. Only for large relaxation times oscillations were found in the solution for temperature. These oscillations were found to be subcritical and damped. The time conditions used by Taitel are unrealistic from the physical realities of heat transfer. That is why their solution exhibited a temperature overshoot. Thus the equations do not violate the laws of thermodynamics as much as the choice of space and time conditions as constraints. Sharma[8] also showed that a temperature undershoot can occur when Fourier's law of heat conduction is applied at steady state in the presence of a temperature dependent heat source. This is in violation of the third law of thermodynamics. Here again, the choice of the space condition at some arbitrary length is not sufficient. The critical length beyond which no heat transfer will occur will have to be identified to keep the solution from violating the third law of thermodynamics.

Jiang [21] reported a solution to the hyperbolic heat equation in hollow spherical objects by the method of integral transformation. Tsai and Hung [22] studied the dynamic behavior of a bi-layered composite sphere subject to sudden change on the outer surface using the Riemann-sum approximation and Laplace transformation. There is no reported study on the effects of non-Fourier heat conduction in TBC especially during a fire. In this study the finite speed conduction effects on the TBC and what are the important design implications are explored.

II. MATHEMATICAL MODEL

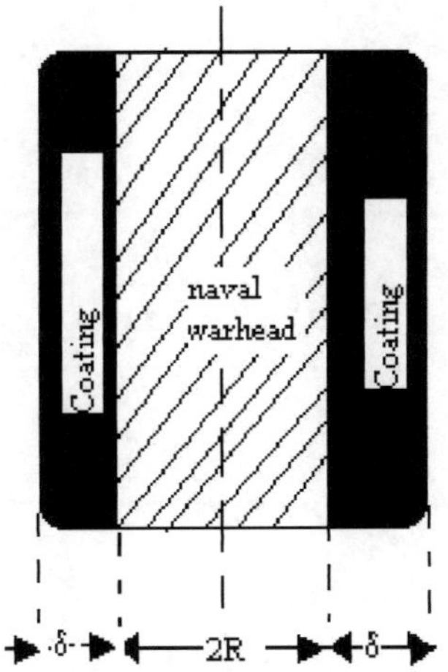

Figure 1.0 Thermal Barrier Coating Applied on Surfaces with High Temperature Application

Let the temperature inside the naval warhead be given by T_n and the temperature inside the coating by given by T_c. The naval warhead and the TBC are idealized into a cylindrical geometry. The radius of the warhead is given by R and the thickness of the coating is represented by δ. The generalized Fourier's law of heat conduction is combined with the energy balance equation in the r direction. The heat propagative velocity is given as the square root of the ratio of thermal diffusivity and relaxation time, $V_{nh} = \sqrt{\frac{\alpha_n}{\tau_{nr}}}$ for propagation in the naval warhead and $V_{ch} = \sqrt{\frac{\alpha_c}{\tau_{cr}}}$. Neglecting any heat sources or sinks the governing equation for the temperatures in the naval warhead and TBC can be written as follows;

$$\tau_{nr}\frac{\partial^2 T_n}{\partial t^2} + \frac{\partial T_n}{\partial t} = \alpha_n \frac{\partial^2 T_n}{\partial r^2} + \frac{\alpha_n}{r}\frac{\partial T_n}{\partial r} \tag{1}$$

$$\tau_{cr}\frac{\partial^2 T_c}{\partial t^2} + \frac{\partial T_c}{\partial t} = \alpha_c \frac{\partial^2 T_c}{\partial r^2} + \frac{\alpha_c}{r}\frac{\partial T_c}{\partial r} \tag{2}$$

The governing equations in the dimensionless form for temperatures in the naval warhead and coating respectively are then;

$$\frac{\partial u_n}{\partial \tau} + \frac{\partial^2 u_n}{\partial \tau^2} = \frac{\partial^2 u_n}{\partial X^2} + \frac{1}{X}\frac{\partial u_n}{\partial X} \tag{3}$$

Where $\tau = t/\tau_{nr}$; $u_n = \frac{(T_n - T_{fire})}{(T_0 - T_{fire})}$; $X = \frac{r}{\sqrt{\alpha_n \tau_{nr}}}$ (4)

$$\frac{\partial u_c}{\partial \tau} + \beta \frac{\partial^2 u_c}{\partial \tau^2} = \gamma \frac{\partial^2 u_c}{\partial X^2} + \frac{\gamma}{X}\frac{\partial u_c}{\partial X} \tag{5}$$

Where $u_c = \frac{(T_c - T_{fire})}{(T_0 - T_{fire})}$; $\beta = \frac{\tau_{cr}}{\tau_{nr}}$; $\gamma = \frac{\alpha_c}{\alpha_n}$ (6)

The time and space conditions can be written as;

$X = 0,\ \partial u_n/\partial X = 0$ (7)

$X = X_R,\ k_n \partial T_n/\partial r = k_c \partial T_c/\partial r$ (8)

$X = X_R,\ T_n = T_c$ (9)

$X = X_{R+\delta},\ u_c = 0$ (10)

$t=0,\ 0 \le X \le X_{R+\delta},\ u_n = u_c = 1$ (11)

$t=\infty,\ 0 \le X \le X_{R+\delta},\ u_n = u_c = 0$ (12)

The solution to Eq. (3) is obtained by the method of separation of variables. First the damping term is removed by the substitution, $u = e^{-\tau/2} w_n$.

$$-\frac{w_n}{4} + \frac{\partial^2 w_n}{\partial \tau^2} = \frac{\partial^2 w_n}{\partial X^2} + \frac{1}{X}\frac{\partial w_n}{\partial X} \tag{13}$$

The method of separation of variables can be used to obtain the solution of Eq. (13).

Let $w_n = V(\tau)\, \phi\, (X)$ (14)

Plugging Eq. (14) into Eq. (13), and separating the variables that are a function of X only and τ only the following two ordinary differential equations, one in space and another in time is obtained;

$$\frac{d^2\phi}{dX^2}+\frac{1}{X}\frac{d\phi}{dX}+\lambda^2\phi=0 \tag{15}$$

$$\frac{d^2V}{d\tau^2}=\left(\frac{1}{4}-\lambda^2\right)=0 \tag{16}$$

The solution to Eq. (15) can be seen to a Bessel function of the zeroth order and first kind and Bessel function of the zeroth order and second kind.

$$\phi=c_1J_0(\lambda_nX)+c_2Y_0(\lambda_nX) \tag{17}$$

It can be seen that $c_2 = 0$ as the temperature is finite at $r = 0$. From the boundary condition given by Eq. (7);

$$\frac{\lambda_n R}{\sqrt{\alpha_n\tau_{nr}}}=3.8317+(n-1)\pi\,,\qquad \text{n=1,2,3......} \tag{18}$$

The solution for Eq. (16) is the sum of two exponentials in time, one that decays with time and another that grows exponentially with time.

$$V=c_3\exp\left(\tau\sqrt{0.25-\lambda_n^2}\right)+c_4\exp\left(-\tau\sqrt{0.25-\lambda_n^2}\right) \tag{19}$$

The term containing the positive exponential power exponent will drop out as with increasing time the system may be assumed to reach steady state and that the points within the cylinder will always have temperature values less than that at the boundary. From the final condition in time, i.e., at steady state,

$$w_n=ue^{\tau/2} \tag{20}$$

Thus w will have to be zero at infinite time. Thus c_3 in Eq. (19) is found to be zero. The term containing the positive exponential power exponent will drop out as with increasing time the system may be assumed to reach steady state and the points within the cylinder will always have temperature values less than that at the boundary.

Thus, $$V=c_4\exp\left(-\tau\sqrt{0.25-\lambda_n^2}\right) \tag{21}$$

or, $$u=\sum_0^\infty c_nJ_0(\lambda_nX)\exp\left(\frac{-\tau}{2}-\tau\sqrt{0.25-\lambda_n^2}\right) \tag{22}$$

The c_n can be solved for from the initial condition by using the principle of orthogonality for Bessel functions. At time zero the LHS and RHS are multiplied by $J_0(\lambda_m X)$. Integration between the limits of 0 and R is performed. When n is not m the integral is zero from the principle of orthogonality. Thus when n = m,

$$c_n = \frac{-\int_0^R J_0(\lambda_n X)}{\int_0^R J_0^2(\lambda_n X)} \tag{23}$$

It can be noted from Eq.(34) that when

$$1/4 < \lambda_n^2 \tag{24}$$

the solution will be periodic with respect to time domain. This can be obtained by using De Movries theorem and obtaining the real part to $\exp\left(-i\tau\sqrt{\lambda_n^2 - 0.25}\right)$. Thus for naval warhead materials with relaxation times greater than a certain limiting value the solution for temperature will exhibit subcritical damped oscillations. Thus,

$$\tau_r > \frac{R^2}{15.33\alpha_n} \tag{25}$$

Thus a bifurcated solution is obtained.

$$u = \sum_0^{\infty} c_n J_0(\lambda_n X) \cos\left(\tau\sqrt{\lambda_n^2 - 0.25}\right) \tag{26}$$

Thus the transient temperature profile in the warhead is obtained. For materials with relaxation times greater than $(R^2/15.33\alpha_n)$ where R is the radius of the naval warhead, subcritical damped oscillations can be seen in the transient temperature profile. The exact solution for the naval warhead including the finite speed heat conduction is derived by the method of separation of variables. It is a bifurcated solution. For certain values of λ_n the time portion of the solution is cosinous and damped and for others it is a infinite series of Bessel function of the first kind and 0th order and decaying exponential in time. Also it can be shown that for terms in the infinite series with λ_n is greater than 1/2, the contribution to the solution will be periodic for small R.

The rest of the problem is to solve for the temperature in the coating. The governing equation is given by Eq. (5). Let $u_c = w_c\exp(-m\tau)$. It can be seen that when $m=1/2\beta$, Eq. (5) becomes;

$$\beta\frac{\partial^2 w_c}{\partial\tau^2}-\frac{w_c}{4\beta}=\gamma\frac{\partial^2 w_c}{\partial X^2}+\frac{\gamma}{X}\frac{\partial w_c}{\partial X} \tag{27}$$

Eq. (27) can be solved by the method of separation of variables.

$$\text{Let } w_c = h(\tau)\, g(X) \tag{28}$$

Plugging Eq. (28) into Eq. (27), and separating the variables that are a function of X only and τ only the following two ordinary differential equations, one in space and another in time is obtained;

$$X^2\frac{d^2 g}{dX^2}+X\frac{dg}{dX}+X^2\frac{\lambda_n^2}{\gamma}g=0 \tag{29}$$

$$\frac{d^2 h}{d\tau^2}=\left(\frac{1}{4\beta^2}-\lambda_n^2\right)=0 \tag{30}$$

The solution to Eq. (15) can be seen to a Bessel function of the zeroth order and first kind and Bessel function of the zeroth order and second kind.

$$g=c_1 J_0\left(\frac{\lambda_n r}{\sqrt{\gamma\alpha_n\tau_{nr}}}\right)+c_2 Y_0\left(\frac{\lambda_n r}{\sqrt{\gamma\alpha_n\tau_{nr}}}\right) \tag{31}$$

The solution for Eq. (30) is the sum of two exponentials in time, one that decays with time and another that grows exponentially with time.

$$V=c_3\exp\left(\tau\sqrt{\frac{0.25}{\beta^2}--\lambda_n^2}\right)+c_4\exp\left(-\tau\sqrt{\frac{0.25}{\beta^2}-\lambda_n^2}\right) \tag{32}$$

The term containing the positive exponential power exponent will drop out as with increasing time the system may be assumed to reach steady state and that the points within the cylinder will always have temperature values less than that at the boundary. From the final condition in time, i.e., at steady state,

$$w_c = ue^{\tau/2} \tag{33}$$

Thus w_c will have to be zero at infinite time. Thus c_3 in Eq. (32) is found to be zero. The term containing the positive exponential power exponent will drop out as with increasing time the system may be assumed to reach steady state and the points within the cylinder will always have temperature values less than that at the boundary.

Thus,
$$V = c_4 \exp\left(-\tau\sqrt{\frac{0.25}{\beta^2} - \lambda_n^2}\right) \tag{34}$$

When $\lambda_n > 1/2\beta$, Eq.(34) would become;

$$V = c_4 \cos\left(-\tau\sqrt{\frac{0.25}{\beta^2} - \lambda_n^2}\right) \tag{35}$$

The integration constants in Eq. (31) can be solved for as follows. The boundary condition given by Eq. (8) yield one equation and the boundary condition given by Eq. (9) yields another equation in terms of the integration constants, c_1 and c_2. These two simultaneous equations and simultaneous unknowns can be solved for. The eigen values λ_n can be solved for from the boundary condition given in Eq. (10).

Thus,

$$c_1 J_0\left(\frac{\lambda_n(R+\delta)}{\sqrt{\gamma\alpha_n\tau_{nr}}}\right) + c_2 Y_0\left(\frac{\lambda_n(R+\delta)}{\sqrt{\gamma\alpha_n\tau_{nr}}}\right) = 0 \tag{36}$$

For large values of the argument in the Bessel function, the asymptotic approximation of the function can be used. Hence,

$$c_1 \frac{\sqrt[4]{\gamma\alpha_n\tau_{nr}}(2)}{\sqrt{\pi\lambda_n(R+\delta)\lambda_n(R+\delta)}}\cos\left(\frac{\lambda_n(R+\delta)}{\sqrt{\gamma\alpha_n\tau_{nr}}} - \pi/4\right) + c_2 \frac{\sqrt[4]{\gamma\alpha_n\tau_{nr}}(2)}{\sqrt{\pi\lambda_n(R+\delta)\lambda_n(R+\delta)}}\sin\left(\frac{\lambda_n(R+\delta)}{\sqrt{\gamma\alpha_n\tau_{nr}}} - \pi/4\right) = 0 \tag{37}$$

Or
$$\tan\left(\frac{\lambda_n(R+\delta)}{\sqrt{\gamma\alpha_n\tau_{nr}}} - \pi/4\right) = -\frac{c_2}{c_1} \tag{38}$$

and $$\frac{\lambda_n(R+\delta)}{\sqrt{\gamma\alpha_n\tau_{nr}}} = \tan^{-1}(\frac{c_2}{c_1}) + (2n+1/4)\pi\,, \qquad n=0,1,2,3.....(39)$$

The critical thickness of the coating can be seen to be;

$$\delta < 2\frac{\tau_{cr}}{\tau_{nr}}\sqrt{\gamma\alpha_{nr}\tau_{nr}}(\tan^{-1}(\frac{c_2}{c_1} + (2n+\frac{1}{4})\pi)) - R \tag{40}$$

When the coating thickness is less than the critical thickness given by Eq. (40) subcritical damped oscillations in temperature can be seen. The c_4 can be solved for from the initial time condition by using the principle of orthogonality as done in the case of the naval warhead.

III. CONCLUSIONS

The analytical solution for temperature profile in a naval warhead coated with a thermal barrier coating, TBC is derived for the both the naval warhead and the coating by the method of separation of variables. The external fire temperature, T_{fire} drives the problem. The final condition in time is used as the fourth constraint and second time constraint to obtain bounded and meaningful solutions. These solutions no not disobey the Claudius's Inequality, The solutions were obtained by the method of separation of variables. A critical coating thickness was derived, below which the transient temperature in the coating will undergo subcritical damped oscillations. This shall drive oscillations in temperature within the naval warhead. In the literature coatings are usually by design are made "thin" as possible. The results of this study clearly show that they cannot be made too thin. A desirable mode of operation is when the transient temperature does not oscillate and hence the coating thickness has to be greater than the critical thickness of coating as predicted in Eq. (40). In a similar manner, the naval warhead cannot be made of materials with relaxation time greater than, $R^2/15.33\alpha_{nr}$ the temperature in the naval warhead will undergo subcritical damped oscillations in time. The exact solutions for transient temperatures within the naval warhead and in the TBC are seen to be bifurcated. Both solutions are given by an infinite Bessel series of the first kind and zeroth order and decaying exponential in time. Under some conditions the time domain portion of the solution becomes cosinous.

IV REFERENCES

[1]Ellis, H., *Fire Barrier Coating Composition Containing Magnesium Oxychlorides and High Alumina Calcium ALuminate Cements or Magnesium Oxysulphate,* Delhi Research, **US Patent** 5,572,862, (1986).

[2]UL 1709 *Standard for Safety Rapid Rise Fire Tests of Protection Materials for Structural* Steel, Underwriters Laboratories, ASTM, New York.

[3]Sharma, K. R., *Damped Wave Transport and Relaxation*, Elsevier, Amsterdam, Netherlands, (2005).

[4] Sharma, K. R., "Damped Wave Conduction and Relaxation in Spherical and Cylindrical Coordinates", *Journal of Thermophysics and Heat Transfer*, **Vol. 21**, 4, (2007), 688-693.

[5]Sharma, K. R., "Manifestation of Acceleration during Transient Heat Condution", *Journal of Thermophysics and Heat Transfer,* **Vol. 20,** 4, (2006), 799-808.

[6]Sharma, K. R., "Analytical Solution of Damped Wave Conduction and Relaxation Equation for a finite Sphere and Cylinder", *Journal of Thermophysics and Heat Transfer*, (accepted)

[7]Sharma, K. R., "Temperature Solution in Semi-Infinite Medium under CWT for Cattaneo and Vernotte non-Fourier Heat Conduction", *225th ACS National Meeting*, New Orleans, LA, March 23rd – March 28th , 2003.

[8]Sharma, K. R. "On the Use of Final Condition in Time to Obtain Solution of Cattaneo and Vernotte Damped Wave Transport and Relaxation Equation", *231st ACS National Meeting*, Atlanta, GA, March 26th – March 30th, 2006.

[9]Maxwell, J. C., "On the Dynamical theory of Gases", *Phil. Trans. Roy. Soc.*, **Vol. 157**, (1867), 49.

[10] Morse, P. M. and Feschbach, H., *Methods of Theoretical Physics*, McGraw Hill, New York, 1953.

[11] Cattaneo, C., *Sulla Conduzione del Calore*, Societta Tipografica Modenere, 1948.

[12] Vernotte, P., "Les Paradoxes de la Theorie Continue de l'equation de la Chaleur", *C. R. Hebd. Seanc. Acad. Sci. Paris*, **Vol. 246,** 22, (1958), 3154-3155.

[13] Joseph, D. D. and Preziosi, L., "Heat Waves", *Review of Modern Physics*, **Vol. 61**, (1989), 41-73.

[14] Ozisik, M. N., and Tzou, D. Y., :On the Wave Theory of Heat Conduction", *ASME Journal of Heat Transfer*, **Vol. 116**, (1994), 526-535.

[15] Tzou, D. Y., *Macro to Microscale Heat Transfer: The Lagging Behavior*, CRC Press, New York, 1996.

[16] Mitra, K., Kumar, S., Vedavarz, A. and Moallemi, M. K., "Experimental Evidence of Hyperbolic Heat Conduction in Processed Meat", *Journal of Heat Transfer*, **Vol. 117,** (1995), 568-573.

[17] Bai, C. and Lavine, A. S., "On Hyperbolic Heat Conduction and Second Law of Thermodynamics", *Journal of Heat Transfer*, **Vol. 117**, (1995), 256-263.

[18] Taitel, Y., "On the Parabolic, Hyperbolic and Discrete Formulation of Heat Conduction Equation", *International Journal of Heat and Mass Transfer*, **Vol. 15**, 2, (1972), 369-371.

[19] Zanchini, E., "Hyperbolic Heat Conduction Theories and non-Decreasing Entropy", *Phys. Review B – Condensed Matter and Material Physics*, **Vol. 60**, 2, (1999), 991-997.

[20] Barletta, A. and Zanchini, E., "Thermal-Wave Heat Conduction in a Solid Cylinder which undergoes a Change of Boundary Temperature", *Heat and Mass Transfer/Warema-und Stoffuebertragung*, **Vol. 32**, 8, (2003), 53583-5386.

[21] Jiang, F., "Solution and Analysis of Hyperbolic Heat Propagation in Hollow Spherical Objects", *Heat and Mass Transfer*, **Vol. 42,** 12, (2006), 1083-1096.

[22] Tsai, C. S. and Hung, C. I., "Thermal Wave Propagation in Bi-Layered Composite Sphere due to a Sudden Temperature Change on the Outer Surface", *International Journal of Heat and Mass Transfer*, **Vol. 46,** (2003), 5137-5144,

TRANSIENT DAMPED WAVE CONDUCTION AND RELAXATION IN HUMAN SKIN AND THERMAL WEAR DURING WINTER

Kal Renganathan Sharma[+]
Prairie View A & M University
Prairie View, TX 77446

ABSTRACT

Damped wave conduction and relaxation in the human skin layer and thermal fabric layer are modeled with a temperature dependent heat source in the human tissue layer. Steady state temperature profiles are derived from the Fourier heat conduction equation. The general solution for the temperature is assumed to be a sum of the transient temperature and steady state temperature. This makes the boundary conditions in space for the skin and fabric layers homogeneous for the transient temperarature. The hyperbolic PDE is solved for by the method of separation of variables. The use of final condition in time in addition to the initial temperature condition leads to bounded infinite Fourier series solutions. These solutions are bounded and does not violate second law of thermodynamics. The model can be used to interpret experimental observations of maximum heat flux that is a parameter of the warm/cool feeling of human skin in winter. For large relaxation times of human skin tissue, $\tau_{rs} > \frac{(1+U^*)^2(b-a)^2}{16\pi^2\alpha_s}$, *the transient temperature can be expected to undergo oscillations. These oscillations will be supercritical and grow with time for strong heat sources,* $U^* > 1$ *and may be subcritical damped oscillatory for weak heat sources,* $U^* < 1$. *For large thermal relaxation times of thermal fabric material,* $\tau_{rf} > \frac{a^2}{4\pi^2\alpha_s}$, *the transient temperature in the thermal fabric layer may be expected to be subcritical damped oscillatory.*

NOMENCLATURE

a	thickness of thermal fabric layer (m)
b	thickness of thermal fabric layer + thicknes of human skin tissue (m)
C_{ps}	heat capacity of human skin tissue (J/Kg/K)
C_{pf}	heat capacity of thermal fabric layer (J/kg/K)
k_f	thermal conductivity of thermal fabric material (w/m/K)
k_s	thermal conductivity of human skin tissue (w/m/K)
m,n	index of eigen values in Fourier infinite series,
q	heat flux (w/m^2)

[+] Adjunct Professor, Department of Chemical Engineering, Room 201A C. L. Wilson Building, PO Box 519, MS 2505. Email: jyoti_kalpika@yahoo.com

q*	dimensionless heat flux ($q^* = \dfrac{q\sqrt{\tau_{rs}}}{(T_{bl} - T_{amb})\sqrt{k_s \rho_s C_{ps}}}$)
S	storage coefficient ($\rho C_p/\tau_r$) (w/m^3/K)
T_{amb}	ambient temperature (K)
T_{bl}	blood temperature (K)
u	dimensionless temperature ($\dfrac{(T - T_{amb})}{(T_{bl} - T_{amb})}$
u^t	transient dimensionless temperature
u^s	steady state dimensionless temperature
U*	dimensionless heat source (U'''/S)
U'''	temperature dependent heat source (w/m^3/K)
w	wave temperature, u = exp(-nτ)
X	dimensionless distance ($\dfrac{x}{\sqrt{\alpha_s \tau_{rs}}}$)

Greek

α_f	thermal diffusivity of thermal fabric material (m^2/s)
α_s	thermal diffusivity of human skin tissue (m^2/s)
β	ratio of thermal diffusivity of skin layer to fabric layer(α_s/α_f)
γ	ratio of thermal relaxation time of skin layer to fabric layer (τ_{rs}/τ_{rf})
λ_n, λ_m	eigen values
ρ_f	density of thermal fabric material (kg/m^3)
ρ_s	density of human skin tissue (kg/m^3)
τ_{rs}	thermal relaxation time of human skin tissue (s)
τ_{rf}	thermal relaxation time of thermal fabric layer (s)
τ	dimensionless time (t/τ_{rs})

I. INTRODUCTION

The sensation of warmth/coolness to contact with human skin with thermal fabric is a problem in transient heat conduction. It is causative to the perception of comfort level of a garment. The heat transfer between human skin and the thermal wear in contact with the human skin is the cause of the warm/cool feeling felt by the human skin. Uchida[1] attempted to correlate between the warm/cool feeling and the physical properties of the thermal wear or the object in contact with the human skin. Yoneda and Kawabata[2] developed an apparatus to measure the transient heat conduction within a short contact time after contact of fabric using their apparatus. Theoretical analysis of the phenomena has largely been analysis of Fourier transient conduction equation with time varying boundary condition[3]. They indicated that the maximum heat flux was a convenient measure of warm/cool feeling of the fabric. They obtained a high correlation between physically measured maximum heat flux and fabric warm/cool feeling data generated in human sensory tests. A high maximum heat flux corresponded to the cool feeling and a low maximum heat flux corresponds to warm feeling. They found that the maximum heat flux depends on the fabric surface and not on the number of fabric layers or fabric thickness. They also found that the maximum heat flux was sensitive to fabric water content and surface geometry.

Considering the short time scales associated with the phenomena and in light of the recently reported experimental values of thermal relaxation times of biological material by Mitra and Kumar et. al.[4] the damped wave conduction and relaxation may be an important consideration in interpretation of the phenomena of sensational perception of human skin. Sharma[5] has discussed the manifestation of damped wave condition and relaxation in a number of transient applications.

Six reasons were given by Sharma[6] to seek a generalized Fourier's Law of Heat Conduction. The generalized Fourier's law of heat conduction can be combined with an energy balance equation, made dimensionless and written for 1 dimension in Cartesian coordinates as;

$$\frac{\partial u}{\partial \tau} + \frac{\partial^2 u}{\partial \tau^2} = \frac{\partial^2 u}{\partial X^2} \qquad (1)$$

This is the damped wave conduction and relaxation equation. When the relaxation time is zero, Eq. (1) will revert to the parabolic partial differential equation, PDE, for transient conduction from the Fourier's law. When the rate of change of temperature with time is much greater than an exponential rise with time, ,e^{τ}, Eq. (1) will revert to the wave equation, (Tzou[7], Sharma[5]). Eq .(1) is a hyperbolic partial differential equation, that is second order with respect to space and second order with respect to time. Reference to the generalized Fourier's law of heat conduction can be traced back to Maxwell [8], Morse and Feshbach [9]. Cattaneo [10] and Vernotte [11] postulated this equation independently. This equation can be used to account for finite speed of heat . Reviews in the literature of the use of Eq. (1) have been provided by Joseph and Preziosi [12] and Ozisik and Tzou [13]. Micro to macro scale behavior has been discussed by Tzou[13]. Sharma [5] discussed the manifestation of the damped wave transport and relaxation equation in industrial applications and provided bounded solutions within the constraints of second law of thermodynamics. It was shown that the generalized Fourier's law of heat conduction can be derived by including the acceleration term in the free electron theory, the acceleration term in the Stokes-Einstein theory for molecular diffusion, by accounting for the accumulation term in the kinetic theory of gases and combining in series the Hooke's elastic element and Newton's viscous element in the viscoelastic theory. The relaxation time was found to be a third of the collision time of the electron and the obstacle. The velocity of heat was found to be identical with the velocity of mass derived from kinetic representation of pressure or the Maxwell representation of the speed of molecules. Some investigators [14] have used the Boltzmann transport equation and derived both the Fourier's law of heat conduction and the damped wave conduction and relaxation equation as special cases. They derive a set of equations for length scales comparable to the mean free path of the molecule. Ali [15,16] used statistical mechanics and the kinetic theory and derived the generalized Fourier's law of heat conduction for monatomic and diatomic gases. Glass and McRae [17] looked at the variable specific heat and thermal relaxation parameter. Luikov [18] discussed the hyperbolic heat conduction equation. He provided a range of relaxation times from milliseconds to 1000 seconds. Both heat transfer in turbulent systems and heat conduction in metals can be represented using the same relaxation parameter.

The relaxation time has been measured by Brown and Churchill [19], Peshkov [20], Zehnder and Rosakis [21]. The relaxation mechanism is fundamental to thermal resonance that cannot be depicted by Fourier's law of heat conduction [22]. For the thermal wave speed around 900 m/s in 4340 steel at 480 ^{0} C the value of relaxation time was found to be of the order of 10^{-11} seconds. A table for relaxation time for different materials at different temperatures and pressures are not available in the literature. Relaxation times for materials with a non-homogeneous inner structure were presented by Kaminski [23]. For sodium bicarbonate they report a relaxation time of 29 seconds and 20 seconds for sand and 54 seconds for ion exchange materials. Mitura et. al. [24] claim that for the falling drying rate period the average time is of the order of several thousand seconds. For homogeneous substances the relaxation time values range from 10^{-8} to 10^{-10} seconds for gases, 10^{-10} – 10^{-12} seconds for liquids and dielectric solids as concluded by Sieniutycz [25]. Mitra and Kumar et. al. [4] presented experimental evidence of the wave nature of heat propagation in processed meat and demonstrated that the hyperbolic heat conduction model is accurate representation on a macroscopic level of the heat conduction process in such biological material. They report a relaxation time of the order of 16 seconds.

Some investigators have raised some concerns about violations of second law of thermodynamics by the generalized Fourier's law of heat conduction (Bai and Lavine [26], Taitel [27] Barletta and Zanchini [28]). Taitel [29] attempted to obtain analytical solution to the governing equation and found that the solution temperature for some values exceeded the boundary temperature indicating a possible violation of Clausius inequality. Al Nimr et. al.[30] discusses the 'temperature overshoot' phenomena. Haji Sheik et al[31] pointed out some anomalies in the hyperbolic heat equation. Transient instability, including the intrinsic transition from the desirable stability, neutral stability to the ultimate unstable response was investigated by Tzou[32] for a wide spectrum of heating rates. Tzou confirmed that the relaxation time results from the rate equation within the mainframe of the second law in the nonequilibrium, irreversible thermodynamics. Schnaid[33] attempted to derive the governing equation for heat conduction with finite speed of heat propagation directly from classical thermodynamics. Cai et. al[34] presented algebraically explicitly analytical solutions of hyperbolic type heat conduction equations in three dimensions. Antaki[35] examined the dual phase lag equation that was introduced by Tzou and provided an analytical solution for the case of a semi-infinite medium subject to constant wall flux boundary condition. Lewandowsha and Malinowski[36] attempted to provide an analytical solution of the hyperbolic heat conduction equation for the case of finite medium symmetrically heated on both sides using the method of Laplace transforms. Temperature overshoot and undershoot was discussed by Tan and Yang [37] during thermal propagation of thermal waves in a thin film under transient conditions.

Sharma[5] presented an analytical solution for the case of a finite slab subject to constant wall temperature. The final condition in time as the fourth condition for the second order hyperbolic PDE governing equation was shown to result in well bounded solutions. This means that care must go into the choice of the conditions used in the boundaries of space and initial and final time values. They have to be physically realistic such as at steady state equilibrium temperature is attained. Only for large relaxation times oscillations were found in the solution for temperature. These oscillations were found to be subcritical and damped. The time conditions used by Taitel are unrealistic from the physical realities of heat transfer. That is why their solution exhibited a temperature overshoot. Thus the equations do not violate the laws of thermodynamics as much as the choice of space and time conditions as constraints. Sharma[5] also showed that a temperature undershoots can occur when Fourier's law of heat conduction is applied at steady state in the presence of a temperature dependent heat source. This is in violation of the third law of thermodynamics. Here again, the choice of the space condition at some arbitrary length is not sufficient. The critical length beyond which no heat transfer will occur will have to be identified to keep the solution from violating the third law of thermodynamics.

There is no reported work on the effect of damped wave conduction and relaxation on warm/cool feeling of the human skin. In this study a two layer mathematical model is developed to study the transient heat conduction of the human skin and thermal fabric layer during use to protect the human body from cold winter outdoors.

II. THEORY

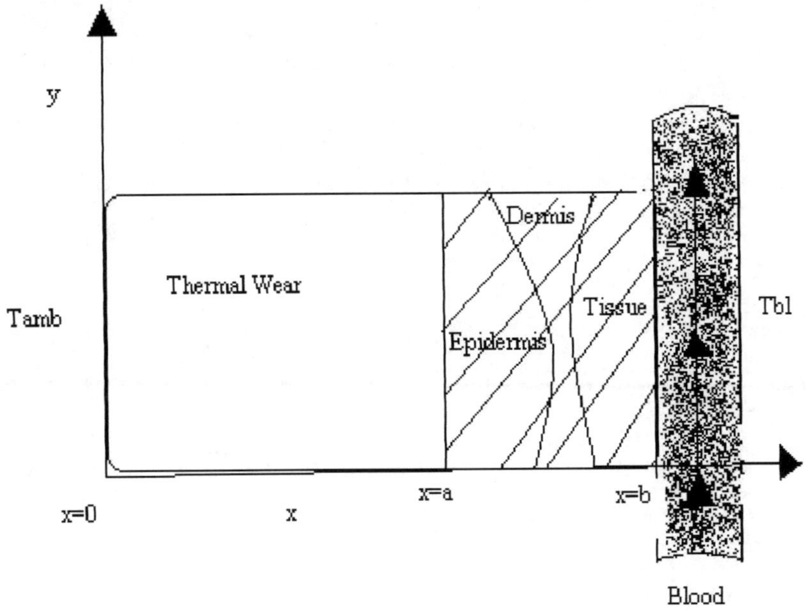

Figure 1-0 Transient Heat Conduction in Human Skin and Thermal Wear in Winter

A schematic of the relevant aspects from a cross-section of human skin near a finger pad is shown in Figure 1-0. Many kinds of receptors in human skin are known to transmit information about the surroundings to the central nervous system. The role of these receptors in generating sensations caused by stimulus from the surroundings is analyzed. The response in the receptors is physico-chemical in nature. The correlation between mechanical stimulus and sensation of touch from a neurophysiological standpoint was studied experimentally by La Motte et. al[38]. They discuss the relation between the surface roughness of fabric and human sensory feel. "Krause's end bulb" is the receptor attributed with the cool feeling and "Ruffini's ending" is the receptor that is responsible for warm feeling. The transient heat conduction in the neighborhood of these receptors as the outside temperature plummets to low levels as typical of winter in the northern part of USA needs to be modeled. The skin layer and the thermal layer assembly are approximated as shown in Figure 1-0. The blood flow in the vessels results in a constant temperature environment at x = b at $T = T_{bl}$, where T_{bl} is the blood temperature. The origin is taken at the interface between the winter surroundings and outer surface of the thermal wear used to protect the skin from the winter cold weather. The thickness of the thermal fabric is 'a' and the interface of the thermal wear and human skin occurs at x = a. Let the ambient temperature be T_{amb}. In winter this can be expected to be much lower than the blood temperature, i.e., $T_{amb} << T_{bl}$. The temperature difference between the blood vessels and the cold surroundings drives the problem. Let the dimensionless temperature time and penetration distance be defined with respect to the human skin parameters as follows;

$$u = \frac{T - T_{amb}}{T_{bl} - T_{amb}}; \qquad \tau = \frac{t}{\tau_{rs}}; \qquad X = \frac{x}{\sqrt{\alpha_s \tau_{rs}}} \tag{2}$$

A. STEADY STATE

At steady state the governing equation for temperature in the skin layer can be written as follows. A heat source in the skin layer generates heat to provide the warm feeling that comes in winter. Let this heat source be dependent on temperature difference with the ambient temperature. The energy balance on a thin slice with thickness Δx and cross-sectional area A can be written as;

$$-\frac{\partial q_x}{\partial x} + U'''(T - T_{amb}) = 0 \qquad (3)$$

Writing the Fourier's law of heat conduction for the heat flux, q_x, in the dimensionless form Eq. (3) may be written as;

$$\frac{\partial^2 u}{\partial X^2} = -U^* u \qquad (4)$$

Where $U^* = \frac{U'''}{S}; S = \frac{\rho_s C_{ps}}{\tau_{rs}}$

The solution for Eq. (4) can be written as;

$$u = A \, Sin(\sqrt{U^*}\, X) + B \, Cos(\sqrt{U^*}\, X) \qquad (5)$$

The boundary conditions at the interface of the skin tissue layer and the blood vessels can be written as;

$$\text{At}, \quad X = X_b = \frac{b}{\sqrt{\alpha_s \tau_{rs}}}, \quad u = 1 \text{ since the } T = T_{bl} \qquad (6)$$

Further it can be assumed that there is no heat transfer from the skin layer to the blood vessels or vice versa. Thus,

$$\text{At}, \quad X = X_b = \frac{b}{\sqrt{\alpha_s \tau_{rs}}}, \quad \frac{\partial u}{\partial X} = 0$$

Applying these boundary conditions given above the integration constants A and B can be solved for and found to be $Sin(\sqrt{U^*}X_b)$ and $Cos(\sqrt{U^*}X_a)$ and Eq. (5) can be seen to be;

$$u = Cos\left[\sqrt{U^*}(X_b - X)\right] \qquad (7)$$

In a similar fashion the governing equation for the temperature in the thermal wear layer, i.e., for $0 \leq x \leq a$, or in terms of dimensionless distance, $0 \leq X \leq X_a$ can be written in the dimensionless form at steady state as follows;

$$\frac{\partial^2 u}{\partial X^2} = 0 \qquad (8)$$

The solution to Eq. (8) can be seen to be;

$$u = CX + D \tag{9}$$

The boundary conditions for the temperature in the thermal wear layer can be written as;

$$X = 0, \;\; u = 0 \tag{10}$$

$$X = X_a, \;\; u\text{ (thermal wear)} = u(\text{skin}) = Cos(X_b - X_a) \tag{11}$$

Thus C and D in Eq. (9) can be solved for using the boundary conditions given by Eq. (10) and Eq. (11). Thus,

$$u = \frac{xCos\left[\sqrt{U^*}(X_b - X_a)\right]}{a} \tag{12}$$

B. TRANSIENT STATE IN HUMAN SKIN LAYER

The energy balance equation in the skin layer during transient state can be written as;

$$-\frac{\partial q}{\partial x} + U'''(T - T_{bl}) = (\rho_s C_{ps})\frac{\partial T}{\partial t} \tag{13}$$

Combining Eq. (13) with the damped wave conduction and relaxation equation given by;

$$q = -k_s \frac{\partial T}{\partial x} - \tau_{rs} \frac{\partial q}{\partial t} \tag{14}$$

and making the terms dimensionless the following governing equation can be written for the skin layer undergoing transient heat conduction.

$$\frac{\partial^2 u}{\partial \tau^2} + \frac{\partial u}{\partial \tau}(1 - U^*) = \frac{\partial^2 u}{\partial X^2} + uU^* \tag{15}$$

where u, τ, U^* and X are defined by Eq. (1) and Eq. (3). It can be seen that the boundary conditions are not homogeneous. Hence the solution is assumed to take the form;

$$u = u^t + u^s \tag{16}$$

where u^t is the transient temperature and u^s = steady state temperature. The solution for u^s was solved for in the case 1 above and given by Eq. (7). The rest of the problem is obtaining the solution of the transient temperature subject to the following time and space conditions.

$$\frac{\partial^2 u^t}{\partial \tau^2} + \frac{\partial u^t}{\partial \tau}(1 - U^*) = \frac{\partial^2 u^t}{\partial X^2} + u^t U^* \tag{17}$$

The initial and final time and space conditions can be written as;

at $\tau = 0$, $u^t = 1$ (18)

at $\tau = \infty$, $u^t = 0$ (19)

at $X = X_b$, $u^t = 0$ (20)

at $X = X_a = u^t = u - u^s$ (21)

Assuming that equilibrium is established rapidly at the interface, $u - u^s$ in the RHS of Eq. (21) can be taken as zero. Hence the boundary condition at $X = X_a$, $u^t = 0$.

The transient temperature, u^t, can be expected to have an exponential decaying component or the damping component in the Eq. (17) can be removed by the following substitution, $u^t = w\exp(-n\tau)$. For $n = (1-U^*)/2$, Eq. (17) is transformed into a governing equation of the wave temperature, w,

$$\frac{\partial^2 w}{\partial \tau^2} - \frac{w(1+U^*)^2}{4} = \frac{\partial^2 w}{\partial X^2} \tag{22}$$

Eq. (22) can be solved for by the method of separation of variables. Let $w = V(\tau)\phi(X)$. Then the terms in Eq. (22) can be separated into two equations, one in space and another in time domain.

$$\frac{\phi''}{\phi} = -\lambda_n^2 = \frac{V''}{V} - \frac{(1+U^*)^2}{4} \tag{23}$$

The general solution in the space domain to the second order ODE with constant coefficient can be seen to be;

$$\phi = c_1 \mathrm{Sin}(\lambda_n X) + c_2 \mathrm{Cos}(\lambda_n X) \tag{24}$$

The boundary conditions given by Eq. (20) and Eq. (21) are applied to Eq. (24). Thus

$$c_2 = -c_1 \mathrm{Tan}(\lambda_n X_b) \tag{25}$$

$$\lambda_n = \frac{2n\pi}{(X_b - X_a)}, \qquad n=1,2,3,...... \tag{26}$$

The time domain portion of the solution to the second order ODE with constant coefficients can be written as;

$$V = c_3 \exp(+\tau\sqrt{\frac{(1+U^*)^2}{4} - \lambda_n^2}) + c_4 \exp(-\tau\sqrt{\frac{(1+U^*)^2}{4} - \lambda_n^2}) \tag{27}$$

From the final condition given by Eq. (19) at $\tau = \infty$, $w = u^t \exp(+n\tau) = 0*\infty=0$. Hence, c_3 can be seen to be zero. The general solution for the transient temperature can be written ;

$$u = \sum_{n=1}^{\infty} c_n e^{-\tau\frac{(1-U^*)}{2}} e^{-\tau\sqrt{\frac{(1+U^*)^2}{4}-\lambda_n^2}} (Tan(\lambda_n X_b)(Sin(\lambda_n X) + Cos(\lambda_n X)) \tag{28}$$

λ_n is given by Eq. (26). c_n can be solved for from the initial condition given by Eq. (18) and using the principle of orthogonality. c_n is found to be ;

$$c_n = \frac{2(1-(-1)^n)(X_b - X_a)}{n^2\pi^2 Tan(\lambda_n X_b)} \tag{29}$$

C. TRANSIENT STATE IN THERMAL FABRIC LAYER

In a similar fashion the transient temperature in the thermal wear layer can be calculated as follows. The governing equation for transient temperature in the thermal fabric layer can be written as;

$$\beta\frac{\partial^2 u}{\partial X^2} = \frac{\partial u}{\partial \tau} + \gamma\frac{\partial^2 u}{\partial \tau^2} \tag{30}$$

Where $\beta = \frac{\alpha_f}{\alpha_s}$; $\gamma = \frac{\tau_{fr}}{\tau_{sr}}$ and u, τ and X are the same as defined in Eq. (1).

It can be seen that the boundary conditions are not homogeneous. Hence the solution is assumed to take the form;

$$u = u^t + u^s \tag{31}$$

where u^t is the transient temperature and u^s = steady state temperature. The solution for u^s was solved for in the section on steady state above and given by Eq. (12). The rest of the problem is obtaining the solution of the transient temperature subject to the following time and space conditions.

$$\gamma \frac{\partial^2 u^t}{\partial \tau^2} + \frac{\partial u^t}{\partial \tau} = \beta \frac{\partial^2 u^t}{\partial X^2} \tag{32}$$

The initial and final time and space conditions can be written as;

$$\text{at } \tau = 0, \quad u^t = 1 \tag{33}$$

$$\text{at } \tau = \infty, \quad u^t = 0 \tag{34}$$

$$\text{at } X = 0, \quad u^t = 0 \tag{35}$$

$$\text{at } X = X_a = \; u^t = u - u^s \tag{36}$$

Assuming that equilibrium is established rapidly at the interface, $u - u^s$ in the RHS of Eq. (36) can be taken as zero. Hence the boundary condition at $X = X_a$, $u^t = 0$.

The transient temperature, u^t, can be expected to have an exponential decaying component or the damping component in the Eq. (32) can be removed by the following substitution, $u^t = w\exp(-n\tau)$. For $n = \frac{1}{2\gamma}$, Eq. (32) is transformed into a governing equation of the wave temperature, w,

$$\gamma \frac{\partial^2 w}{\partial \tau^2} - \frac{w}{4\gamma} = \beta \frac{\partial^2 w}{\partial X^2} \tag{37}$$

Eq. (37) can be solved for by the method of separation of variables. Let $w = \theta(\tau)g(X)$. Then the terms in Eq. (37) can be separated into two equations, one in space and another in time domain.

$$\beta \frac{g''}{g} = -\lambda_m^2 = \gamma \frac{\theta''}{\theta} - \frac{1}{4\gamma} \tag{38}$$

The general solution in the space domain to the second order ODE with constant coefficient can be seen to be;

$$g = c_1 Sin(\lambda_m X) + c_2 Cos(\lambda_m X) \quad (39)$$

The boundary conditions given by Eq. (35) and Eq. (36) are applied to Eq. (39). Thus

$$c_2 = 0 \quad (40)$$

$$\lambda_m = \frac{m\pi}{X_a}, \qquad m=1,2,3,\ldots\ldots \quad (41)$$

The time domain portion of the solution to the second order ODE with constant coefficients can be written as;

$$\theta = c_5 \exp(+\tau\sqrt{\frac{1}{4\gamma^2} - \frac{\lambda_m^2}{\gamma}}) + c_6 \exp(-\tau\sqrt{\frac{1}{4\gamma^2} - \frac{\lambda_m^2}{\gamma}}) \quad (42)$$

From the final condition given by Eq. (34) at $\tau = \infty$, $w = u^t\exp(+n\tau) = 0*\infty=0$. Hence, c_5 can be seen to be zero. The general solution for the transient temperature can be written ;

$$u = \sum_{n=1}^{\infty} d_n e^{\frac{-\tau}{2\gamma}} e^{-\tau\sqrt{\frac{1}{4\gamma^2} - \frac{\lambda_m^2}{\gamma}}} Sin(\lambda_m X) \quad (43)$$

λ_m is given by Eq. (41). d_n can be solved for from the initial condition given by Eq. (33) and using the principle of orthogonality. d_n is found to be ;

$$d_n = \frac{2(1-(-1)^n)}{m^2\pi^2} \qquad , n = 1,2,3,\ldots. \quad (44)$$

IV DISCUSSION

The solution to the transient heat conduction in the human skin layer and the thermal fabric layer including the damped wave conduction and relaxation effects were derived using the method of separation of variables. The use of the final condition in time leads to well bounded, physically realistic solutions within the bounds of Clausius's inequality. The transient temperature in the two layers at steady state is shown in Figure 2-0. The nature of the temperature profile is cosinous in the human skin layer and linear in the thermal fabric layer. The parameters the profile is derived for are, $X_a = 3$, $X_b = 5$ and $U^* = 2.0$. The heat flux at steady state for the human skin layer and the thermal fabric layer is shown in Figure 3-0 for the parameters, $X_a = 3$, $X_b = 5$, $k_f/k_s = 5.0$. and $U^* = 2.0$. It can be seen from Figure 3-0 the interface heat flux undergoes a maxima. For the heat flux in dimensionless form with respect to the skin thermophysical properties to be continuous the ratio of the thermal conductivities of the thermal fabric layer and skin layer have to be a certain value and cannot be specified independently. This can be deduced from Eq. (7) and Eq. (12). The maximum heat flux at the interface of the two layers may be related to the heat flux reported by Yoneda and Kawabata[3].

It can be seen from the model solutions that the transient temperature will undergo subcritical damped oscillations under certain conditions. These conditions are for large relaxation times of the skin and for large relaxation times for the thermal fabric layer. Thus for,

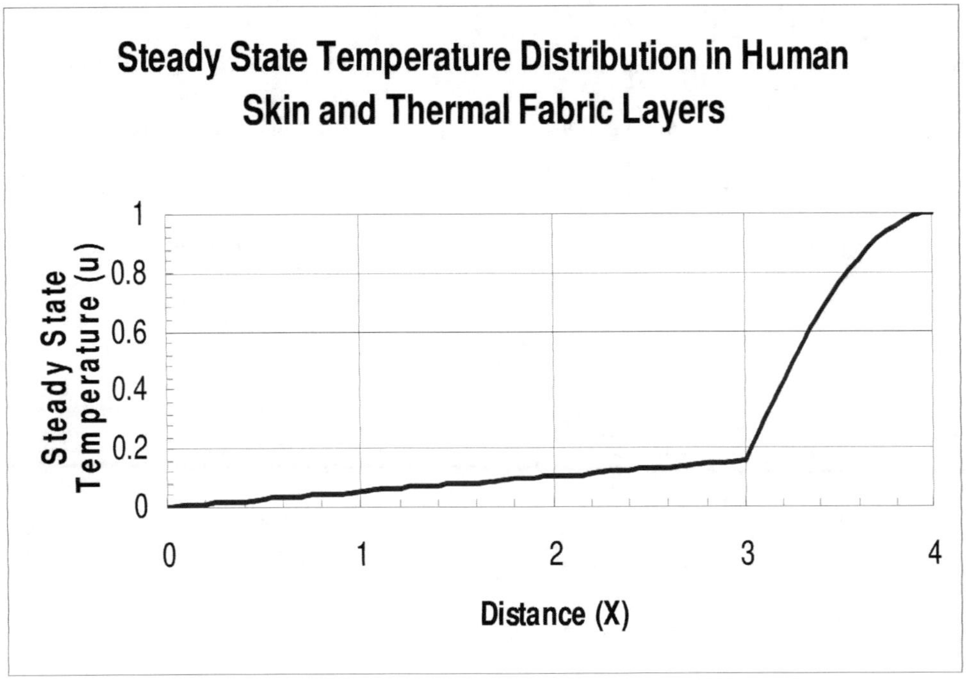

Figure 2-0 Steady State Temperature in Human Skin Layer and Thermal Fabric Layer for $X_a = 3.0$ and $X_b = 4.0$ and $U^* = 2.0$

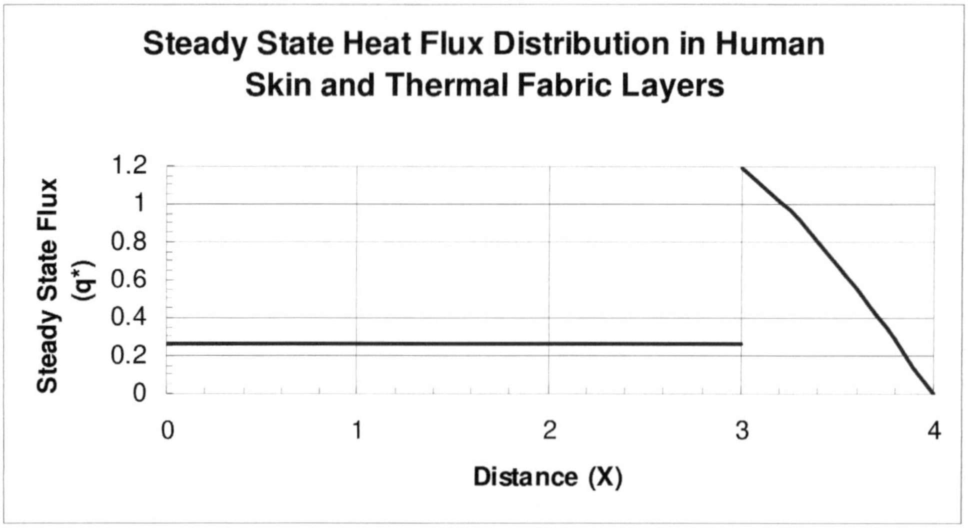

Figure 3-0 Steady State Temperature in Human Skin Layer and Thermal Fabric Layer for $X_a = 3.0$, $X_b = 4$, $k_f/k_s = 5$ and $U^* = 2.0$

Thus for large relaxation time values of the skin, i.e., $\tau_{rs} > \dfrac{(1+U^*)^2(b-a)^2}{16\pi^2\alpha_s}$, it can be seen that the temperature in the skin layer can be expected to exhibit oscillations. The nature of the oscillations will depend on the strength of the heat source which is a function of the temperature difference with the ambient cold winter temperature. For heat source, U* > 1 oscillations that grows with time may be expected. For heat source, U* < 1 subcritical damped oscillations can be expected. The solution for these materials is then given by;

$$u = \sum_{n=1}^{\infty} c_n e^{-\tau\frac{(1-U^*)}{2}} Cos(\tau\sqrt{\lambda_n^2 - \frac{(1+U^*)^2}{4}})(Tan(\lambda_n X_b)(Sin(\lambda_n X) + Cos(\lambda_n X)) \tag{45}$$

Where, λ_n and c_n are given by Eq. (26) and Eq. (29) respectively.

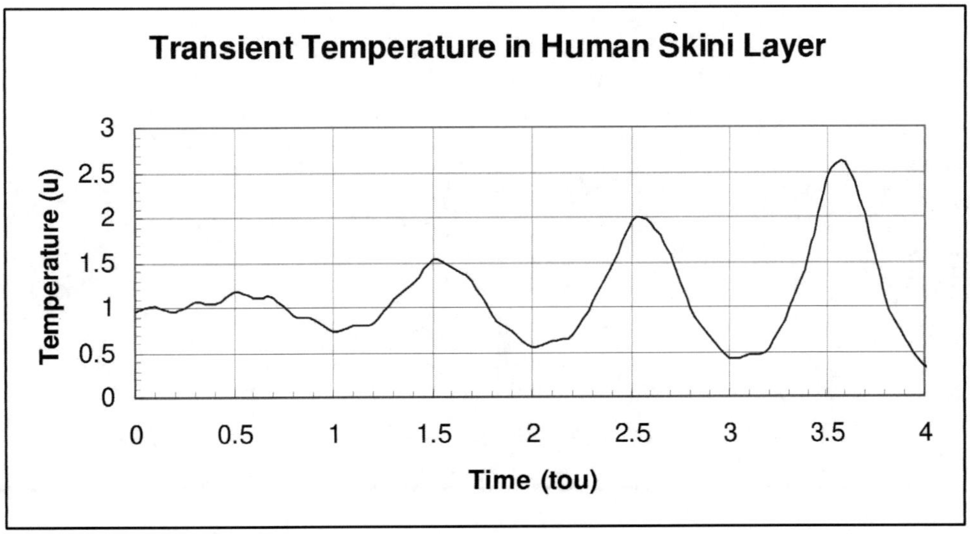

Figure 4-0 Transient Temperature at X = 3.1, in Human Skin Layer for Large Relaxation Time Values, U* = 2.0

Eight terms in the infinite series in Eq. (45) were taken in a Microsoft Excel Spreadsheet and the dimensionless temperature is plotted as a function of dimensionless time in Figure 4-0. It can be seen that for human skin materials with large relaxation times as discussed above the dimensionless temperature exhibits growing oscillations at a heat source of U* = 2.0. The general solution is given by Eq. (45).

In a similar fashion, under certain conditions the temperature in the thermal fabric layer can be expected to undergo subcritical damped oscillations. Thus for,

$$\tau_{rf} > \frac{a^2}{4\pi^2 \alpha_s}$$

the transient temperature in the thermal fabric layer can expect to undergo subcritical damped oscillations. Under these conditions the transient temperature in the thermal fabric layer is given by;

$$u = \sum_{n=1}^{\infty} d_n e^{\frac{-\tau}{2\gamma}} Cos(\tau \sqrt{\frac{\lambda_n^2}{\gamma} - \frac{1}{4\lambda^2}}) Sin(\lambda_m X) \qquad (46)$$

where, d_n and λ_m is given by Eq. (44) and Eq. (41) respectively.

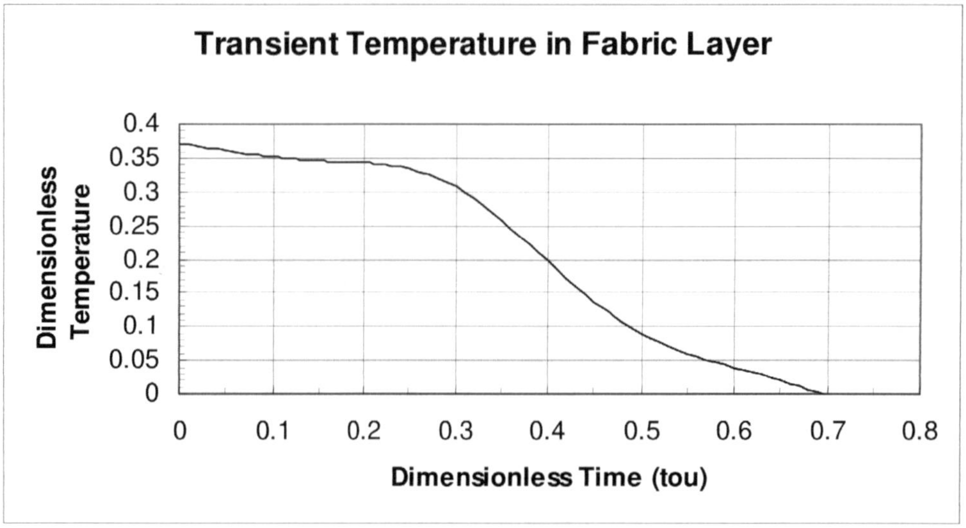

Figure 5-0 Transient Temperature in the Thermal Fabric Layer for Large Relaxation Times at X = 1.1

Six terms in the infinite series given by Eq. (46) were taken and the dimensionless temperature was plotted against dimensionless time in the thermal layer for fabric materials with large relaxation times and shown in Figure 5-0. The γ chosen for the study was 0.15 and the temperature at X = 1.1 was obtained for a X_a = 3.0. It can be seen that the temperature undergoes subcritical damped oscillations.

V CONCLUSIONS

A mathematical model was developed for damped wave conduction and relaxation in the human skin layer and thermal fabric layer using the method of separation of variables. The warm/cool feeling can be explained using the transient temperature and heat flux at the interface. The experimental observations of previous workers on warm/cool feeling such as observation of a maximum heat flux can be explained using the model that has been developed. The Fourier infinite series solutions using the final time condition and appropriate space conditions after subtraction of the steady state temperature to yield homogeneous boundary conditions are seen to be well bounded and within the bounds of Clausius' inequality.

For human skin with large relaxation times, i.e., $\tau_{rs} > \frac{(1+U^*)^2(b-a)^2}{16\pi^2\alpha_s}$, oscillations in temperature that can grow with time can be expected for heat sources with strength greater than 1.0. For heat source strength less than 1 the oscillations can be expected to be subcritical and damped. The solution for the temperature is cosinous with respect to time and given by Eq. (45). For thermal fabric layer with large relaxation times, $\tau_{rf} > \frac{a^2}{4\pi^2\alpha_s}$, subcritical damped oscillations in temperature can be expected. The transient temperature in such cases is given by Eq. (46). The oscillatory behavior in temperature in the human skin layer and the thermal fabric layer are shown as Figures for typical parameters of thickness of fabric, human skin tissue thickness and thermophysical properties of the two layers. The general solution for the transient temperature of the human skin layer, thermal fabric later and the solutions for the steady state temperature of the human skin layer and the thermal fabric layer are presented.

VI REFERENCES

[1] Uchida, H., *Soc. of Fiber Industry*, **Vol. 3,** (1937), 459.

[2] Kawabata, S. and Agaki, Y., *Journal of Text. Mach. Soc. Japan,* **30**, T13, (1977)

[3] Yoneda, M and Kawabata, S., "Analysis of Transient Heat Conduction and Its Applications – Part 2 Theoretical Analysis of the Relationship between Warm/Cool Feeling and Transient Heat Conduction in Skin", *Journal of the Textile Machinery Society of Japan, Trans.*, **Vol. 31,** No. 4, (1985), 79-85.

[4] Mitra, K., Kumar, S., Vedavarz, A. and Moallemi, M. K., "Experimental Evidence of Hyperbolic Heat Conduction in Processed Meat", *Journal of Heat Transfer*, **Vol. 117,** (1995), 568-573.

[5] Sharma, K. R., *Damped Wave Transport and Relaxation*, Elsevier, Amsterdam, 2005.

[6] Sharma, K. R**., "**Damped Wave Conduction and Relaxation in Cylindrical and Spherical Coordinates", *Journal of Thermophysics and Heat Transfer*, **Vol. 21**, 4, (2007), 688-693.

[7] Tzou, D. Y., *Macro to Microscale Heat Transfer: The Lagging Behavior*, CRC Press, New York, 1996.

[8] Maxwell, J. C., "On the Dynamical theory of Gases", *Phil. Trans. Roy. Soc.*, **Vol. 157**, (1867), 49.

[9] Morse, P. M. and Feschbach, H., *Methods of Theoretical Physics*, McGraw Hill, New York, 1953.

[10] Cattaneo, C., " A Form of Heat Conduction Which Eliminates the Paradox of Instantaneous Propagation", *Comptes Rendus Hebdomadaires des Seances de L'academie des Sciences*, **Vol, 247**, (1958), 431-3=433.

[11] Vernotte, P., "Les Paradoxes de la Theorie Continue de l'equation de la Chaleur", *C. R. Hebd. Seanc. Acad. Sci. Paris*, **Vol. 246,** 22, (1958), 3154-3155.

[12] Joseph, D. D. and Preziosi, L., "Heat Waves", *Review of Modern Physics*, **Vol. 61**, (1989), 41-73.

[13] Ozisik, M. N., and Tzou, D. Y., :On the Wave Theory of Heat Conduction", *ASME Journal of Heat Transfer*, **Vol. 116**, (1994), 526-535.

[14] Brush, S. G. (ed. & tr.), Boltzmann, *Lectures on Gas Theory*, U. of California Press, Berkeley, CA, 1964.

[15] Ali, A. H., "Statistical Mechanical Derivation of Cattaneo's Heat Flux Law", *Journal of Thermophysics and Heat Transfer*, **Vol. 13**, 4, (1999), 544-545.

[16] Ali, A. H., "Non-Fourier Heat Flux Law in Diatomic Gases", *Journal of Thermophysics and* Heat Transfer, **Vol. 14**, 2, (2000), 281-283.

[17] Glass, D. E. and McRae, D. S ., "Variable Specific Heat and Thermal Relaxation Parameter in Hyperbolic Heat Conduction", *Journal of Thermophysics and Heat Transfer*, **Vol. 4**, 2, (1990), 252-254.

[18] Luikov, A. V., *Analytical Heat Diffusion Theory*, Academic, New York, 1968.

[19] Brown, M. A . and Churchill. S. W., "Finite-Difference Computation of the Wave Motion Generated in a Gas by a Rapid increase in the Bounding Temperature", *Computers and Chemical Engineering*, **Vol. 23**, 3, (1999), 357-376.

[20] Peshkov, V., "Second Sound in Helium II", *Journal of Physics USSR*, VIII, (1944), 381-386.

[21] Zehnder A. T. and Roaskis, A. J., "On the Temperature Distribution at the Vicinity of Dynamically Propagating Cracks in 4340 Steel", *Journal Mech. Phys. Solids,* **Vol. 39**, (1991), 384-415.

[22] Tzou, D. Y., "Damping and Resonance Characteristics of Thermal Waves", *ASME Journal Appl. Mech.,* **Vol. 39**, (1991), 384-415.

[23] Kaminski, W., "Hyperbolic Heat Conduction Equation for Materials with a Nonhomogeneous Inner Structure", *Journal of Heat Transfer*, **Vol. 112**, (1990), 555-560.

[24] Mitura, E., Michalowski, S. and Kaminski, W., "A Mathematical Model of Convection Drying in the Falling Drying Rate Period", *Drying Technology*, **Vol. 6**, 10, (1988), 113-137.

[25] Sieniutycz, S., "The Variational Principle of Classical Type for Noncouples Non-Stationary Irreversible Transport Processes with Convective Motion and Relaxation", *International Journal of Heat and Mass Transfer*, **Vol. 20**, (1977), 1221-1231.

[26] Bai, C. and Lavine, A. S., "On Hyperbolic Heat Conduction and Second Law of Thermodynamics", *Journal of Heat Transfer*, **Vol. 117**, (1995), 256-263.

[27] Taitel, Y., "On the Parabolic, Hyperbolic and Discrete Formulation of Heat Conduction Equation", *International Journal of Heat and Mass Transfer*, **Vol. 15**, 2, (1972), 369-371.

[28] Barletta, A. and Zanchini, E., "Thermal-Wave Heat Conduction in a Solid Cylinder which undergoes a Change of Boundary Temperature", *Heat and Mass Transfer/Warema-und Stoffuebertragung*, **Vol. 32**, 8, (2003), 53583-5386.

[29] Al-Nimr, M., Naji, M., and Al-Wardat, S., "Overshooting Phenomenon in the Hyperbolic Heat Conduction Model", *Japanese Journal of Applied Physics, Part I: regular Papers and Short Notes & Review Papers*, **Vol. 42,** 8, (2003), 5383-5386.

[30] Haji-Sheik, A.M, Minkowycz, W. J. and Sparrow, E. M., "Certain Anomalies in the Analysis of Hyperbolic Heat Conduction", *Journal of Heat Transfer*, **Vol. 124**, 2, (2002), 307-319.

[31] Tzou, D. Y., "Thermal Control in Solids with Rapid Relaxation", *Journal of Dynamic Systems, Measurement and Control-Trans. of ASME*, **Vol. 125,** 4, **(2003), 563-568.**

[32] Shnaid, I., "Thermodynamically Consistent Description of Heat Conduction with Finite Speed of Heat Propagation", *International Journal of Heat and Mass Transfer*, **Vol. 46,** 20, (2003), 3853-3863.

[33] Cai, R., Gou, C. and Li, H., "Algebraically Explicit Analytical Solutions of Unsteady 3-D Nonlinear non-Fourier (hyperbolic) Heat Conduction", *International Journal of Thermal Sciences*, **Vol. 45**, 9, (2006), 893-896.

[34] Antaki, P. J., "Solution for non-Fourier Dual Phase Lag Heat Conduction in a Semi-infinite Slab with Surface Heat Flux", *International Journal of Heat and Mass Transfer*, **Vol. 41,** 14, (1998), 2253-2258.

[35] Lewandowsha, M. and Malinowski, L., "An Analytical Solution of the Hyperbolic Heat Conduction Equation for the Case of a Finite Medium Symmetrically Heated in Both Sides", *International Communications in Heat and Mass Transfer*, **Vol. 33**, 1, (2006), 61-69.

[36] Tan, A. M. and Yang W. J., "Propagation of Thermal Waves in Transient Heat Conduction in a Thin Film", *Journal of the Franklin Institute*, **Vol. 336**, 1, (1999), 185-197.

[37] Tian, Y. "Theoretical Investigation of Transport and Reflection of Thermal Waves", Ph.D Dissertation, Oklahoma State University, Stillwater, OK, 1995.

[38] LaMotte, R. H., *Clothing Comfort: Interaction of Thermal, Ventilation, Construction and Assesment Factors*, Ann Arbor Science Publishers, (1977), Ann Arbor, MI.

On the Derivation of the Damped Wave Conduction and Relaxation Equation from Stokes-Einstein Formulation

Kal Renganathan Sharma[1]
Shanmugha Arts, Science, Technology & Research Academy, SASTRA University, Thanjavur, Tamil Nadu, India 613402

Abstract

The damped wave conduction and relaxation equation was derived from the Stokes-Einstein formulation for dilute, binary systems for a slowly moving sphere and analogy between mass diffusion and heat conduction. The velocity of mass derived by accounting for the acceleration of the molecule prior to reaching steady state was found to be identical with the root mean square velocity of molecules of a ideal gas derived from the kinetic representation of pressure. A temperature dependent heat source in a fuel rod was considered. Violations of laws of thermodynamics reported in the literature was found to be due to use of inconsistent boundary conditions and not because of the choice of law such as Fourier or non-Fourier etc. A Bessel composite function solution results and the critical length above which the temperature will remain unchanged is derived when $U^* = 1$. This is when, $X \geq$ sqrt($5.7831/U^* + \tau^2$). A steady state temperature profile is derived. The heat flux is obtained for the rod heated by a temperature dependent heat source. A heat flux reversal was found after a said distance in the rod. A critical point of zero transfer was found. The method of separation of variables was used to obtain the general solution for the temperature profile. A bifurcated solution is found. For length smaller than, $l < 2\pi$sqrt($\alpha\tau_r$)/($1 + U^*$), subcritical oscillations in temperature was found.

Nomenclature

a	=	side of cube with gas molecules (m)
A	=	cross-sectional area of the rod (m^2)
A_N	=	Avagadro number (6.023 E23 molecules/mole)
C	=	concentration ($mole/m^3$)
D_{12}	=	binary diffusivity (m^2/s)
f	=	frictional force (N)
I_0	=	modified Bessel function of the zeroth order and first kind
J	=	molar flux ($mole/m^2/s$)
J_0	=	Bessel fundtion of the zeroth order and first kind
k	=	thermal conductivity (W/m/K)
k_B	=	Boltzmann constant (J/K/molecule)
K_0	=	modified Bessel function of the zeroth order and second kind
l	=	length of rod (m)
m	=	mass of molecule (Kg)
n	=	number of molecules per unit volume (m^{-3})
N_m	=	number of molecules
M	=	molecular weight (gm/mole)
P	=	pressure (N/m^2)
q	=	heat flux (W/m^2)
R	=	gas constant (J/mole/K)
R_o	=	radius of the molecule (m)

[1] Professor, School of Chemical and Biotechnology, Jiva Chaitanya, Email: jyoti_kalpika@yahoo.com

S	=	storage coefficient ($W/m^3/K$)
T	=	temperature (K)
t	=	time (s)
u	=	dimensionless temperature
U'''	=	heat source ($w/m^3/K$)
U^*	=	dimensionless heat source (U'''/S)
v_l	=	velocity of the molecule (m/s)
v_m	=	velocity of mass (m/s)
v_x	=	x component velocity of molecule (m/s)
$<v^2>$	=	root mean square velocity of gas molecules (m/s)
V	=	function of time only
X	=	dimensionless distance $(x/(\alpha\tau_r)^{1/2})$
x_l	=	mole fraction of species 1
Y_0	=	Bessel function of the zeroth order and second kind

Greek

α	=	thermal diffusivity (m^2/s)
ρ	=	density (kg/m^3)
τ_r	=	relaxation time (s)
τ	=	dimensionless time (t/τ_r)
μ	=	viscosity of the solvent (Kg/m/s)
μ_l	=	chemical potential (K/molecule)
η	=	transformation variable ($X^2 - \tau^2$)
ϕ	=	function of space only
λ_n	=	Eigen values

I. Introduction

The basis for a generalized Fourier's law of heat conduction was provided from the free electron theory in a earlier study by Sharma[1]. The acceleration regime of the electron prior to it reaching steady state when accounted for manifests itself as the ballistic term in the generalized Fourier's law of heat conduction and relaxation. Only at steady state the equation of motion for the free electron that takes into account the temperature gradient force, oppositional frictional drag force proportional to the velocity of the electron reverts to the Fourier's law of heat conduction. In the transient state the acceleration term needs to be taken into account. This can apparently be achieved by the generalized Fourier's law of heat conduction. The generalized Fourier's law of heat conduction was originally suggested by Maxwell[2] and postulated independently by Cattaneo[3] and Vernotte[4]. There have been some attempts in the literature to derive the generalized Fourier's law of heat conduction. Sharma[5] has shown that Cattaneo and Vernotte, non-Fourier heat conduction equation can be derived from the kinetic representation of heat flux when the accumulation of energy at the surface is not neglected. Ali[6,7] has used statistical mechanics and kinetic theory of gases and attempted to derive the Cattaneo's heat equation for monatomic and diatomic gases. Glass and MacRae[8] looked at the variable specific heat and the thermal relaxation parameter. Nernst[9] suggested that at low temperatures in good thermal conductors heat may have sufficient inertia to give rise to oscillatory discharge. Ozisik and Tzou[10] have attempted to relate the relaxation time to the electron-phonon collisions and the volumetric heat capacities of the electrons and the metal lattice. The relaxation time has been viewed as a time lag between the heat flux vector and the temperature gradient when the response time is short (Bai and Lavine[11]). Propagation of Landau [12] waves in Helium was discussed as a quantum phenomenon. Experimental measured values of relaxation time are of the order of a few microseconds for stainless steel at 400 ^{0}C as reported by Tzou[13] The relaxation times were found to be of the order of 15-20 seconds for materials with a non-homogeneous inner structure by Kaminski[14] and Mitra, Kumar et. al[15]. It is not clear what the property relaxation time of a given material means. Further work is needed to better understand the fundamentals of heat conduction and all the transient events have to be modeled precisely.

In this study another basis for the generalized Fourier's law of heat conduction is explored from the Stokes-Einstein formulation for diffusion in liquids. It is found that the acceleration regime of the molecule can manifest as the ballistic term in the generalized Fick's law of diffusion and by analogy the Fourier's law of heat conduction.
The role of the acceleration regime of the free electron and that of the molecule can be a causative in the discrepancy between experiment and theory largely derived form Fourier models. In Fourier models the acceleration regime is neglected and steady state is assumed to have set in. One of the reasons for the lack of widespread use of the Cattaneo and Vernotte non-Fourier heat equation was the "violations of laws of thermodynamics" (Taitel[16], Barletta and Zanchini[17], Jou et. al.[18],). As discussed in a earlier study when time conditions are properly used at steady state bounded solutions that are within the constraints of the second law of thermodynamics can be obtained. In this study another scenario where the laws of thermodynamics can be violated if adequate thought did not go into the choice of boundary conditions is undertaked. The effect of temperature dependant heat source is explored in a solid rod Sharma[19]). It can be seen that if care is not taken in the selection of boundary conditions and ensuring that they are consistent with each other the laws of thermodynamics will be broken even at steady state when only the Fourier's law of heat conduction is used to model the system. Thus the temperature undershoot from this study and

overshoot from Taitel[16] are more artifacts of inconsistent and unrealistic boundary conditions and not on account of the basis for non-Fourier heat conduction.

II. Derivation of Generalized Fourier's Law of Heat Conduction from Stokes-Einstein Formulation

The Fick's law of diffusion can be written as;

$$J = -D\partial C/\partial x \quad (1)$$

The generalized Fick's law of diffusion can be written as;

$$J = -D\partial C/\partial x - \tau_r \partial J/\partial t \quad (2)$$

Eq. (2) is written by analogy with the generalized Fourier's law of heat conduction and relaxation (Sharma[10]). The Stokes-Einstein equation can be used to calculate diffusion coefficients in liquids:

$$D = k_B T/f = k_B T/6\pi\mu R_o \quad (3)$$

Eq. (1) can be derived as follows. A rigid solute sphere is assumed for the molecule diffusing in a common solvent. The frictional drag force acting on the molecule opposing its motion is proportional to the velocity of the sphere;

$$\text{Drag force} = f\, v_1 \quad (4)$$

Where f is the friction coefficient. From Stokes law[20] for a sphere moving in a fluid, $f = 6\pi\mu R_o$. The driving force was taken by Einstein[21] to be the negative of the chemical potential gradient defined per molecule.

$$-\nabla\mu_1 = (6\pi\mu R_o)v_1 \quad (5)$$

Eq. (3) is valid when the molecule reaches a steady state velocity. This is when the net force acting on the molecule is zero. Solution is assumed to be ideal and dilute.

$$\mu_1 = \mu_1^0 + k_B T \ln(x_1) = \mu_1^0 + k_B T \ln c_1 - k_B T \ln c_2 \quad (6)$$

For dilute solutions, c_2 far exceeds the solute concentration and can be taken as constant. The gradient at constant temperature, then is;

$$\nabla\mu_1 = k_B T\ \nabla c_1/c_1 = -(6\pi\mu R_o)v_1 \quad (7)$$

$$\text{or } -k_B T/(6\pi\mu R_o)\nabla c_1 = c_1 v_1 = J'' \quad (8)$$

Comparing Eq. (6) with the Fick's law of diffusion given in Eq. (1), the Stokes-Einstein relationship of Eq. (1) results. It can be seen that Eq. (8) is valid only at steady state. Often times, in transient applications a sudden step-change in concentration, i.e., the driving force is imposed on the system. The molecule will experience a accelerating regime prior to reaching steady state. During the accelerating regime;

$$-\nabla\mu_1 - (6\pi\mu R_o)v_1 = m\,dv_1/dt \tag{9}$$

$$\text{or } k_BT\nabla c_1 + mc_1 dv_1/dt = -(6\pi\mu R_o)c_1v_1 \tag{10}$$

$$\text{or } -(k_BT/6\pi\mu R_o)\nabla c_1 = m/6\pi\mu R_o\, \partial J''/\partial t + J'' \tag{11}$$

By analogy the generalized Fourier's law of heat conduction and relaxation can be written as;

$$q = -k\,\partial T/\partial x - \tau_r\, \partial q/\partial t \tag{12}$$

Eq. (10) can be written from Eq. (11) given that a slowly moving sphere is considered (i.e., vdc/dt = 0, Sharma[22]). Further for 1 molecule the concentration is not going to vary in its vicinity to a considerable measure for the term vdc/dt to be significant. Eq. (11) is the generalized Fick's law of diffusion that accounts for the acceleration regime of the molecule as well as the steady state regime. The relaxation time is then given by;

$$\tau_r = m/6\pi\mu R_o = mD/k_BT \tag{13}$$

In terms of P_{tot} the system pressure for ideal gas the relaxation time can be written as;

$$\tau_r = MD\rho/P \tag{14}$$

The velocity of mass diffusion is given by;

$$v_m = (D/\tau_r)^{1/2} = (k_BT/m)^{1/2} \tag{15}$$

Eq. (14) can be rewritten in terms of the molar gas constant and molecular weight as;

$$v_m = (D/\tau_r)^{1/2} = (RT/M)^{1/2} \tag{16}$$

From the Maxwell distribution of molecular speeds[23,24] the average velocity of the molecule is given by $(8RT/M)^{1/2}$. The velocity of mass diffusion is off by a factor of 8 from the average velocity of a molecule in random motion. The kinetic representation of pressure[25] can be written after observing that a molecule moving in a cube of dimensions a with a velocity of v_x undergoes a momentum change of $2mv_x$ upon one collision with the wall. The number of collisions on the wall can be estimated by first calculating the time taken by the molecule to move the roundtrip from the wall after a collision to the opposite wall and return as $2a/v_x$. The number of collisions undergone by a molecule is $v_x/2a$. The rate of transfer of momentum to the surface from the molecular collisions is then, mv_x^2/a. The total force exerted by all the molecules colliding can be obtained b summing the contribution from each molecule and the pressure is obtained by dividing the sum by the area of the wall and is given by;

$$P_{tot} = m/a^3 (v_{x1}^2 + v_{x2}^2 + v_{x3}^2 + \ldots\ldots] \quad (17)$$

Let N_m be the number of molecules in the system and n the number of molecules per unit volume. Then Eq. (16) can be rewritten after multiplying the numerator and denominator by N_m:

$$P_{tot} = mn <v_x^2> = \rho <v_x^2> = 1/3\rho <v^2> \quad (18)$$

As the molecules treated as particles move in random there is no preferred direction in the box. Hence $v^2 = v_x^2 + v_y^2 + v_z^2$. The square root of v^2 is called the root mean squared speed of the molecule and is an widely accepted average molecular speed. From the ideal gas law $P_{tot} = \rho RT/M$. Combining this with Eq. (15);

$$1/3 <v^2> = RT/M = A_N k_B T/M \quad (19)$$

Comparing Eqs. (18) and Eq. (15) it can be seen the velocity of mass is 1/3 of the root mean square velocity. This could be due to the one dimensional diffusion.

III. Temperature Dependent Heat Source

Consider a rod of length l, with one end maintained at temperature T_s. The other end is at the zero temperature (O K). This is the lowest temperature achievable according to the third law of thermodynamics. The entropy is zero at 0 K. At time t = 0, the rod is at 0 K. For times greater than zero, the temperature dependent heat source is allowed to heat the rod. It is of interest to study the temperature distribution in the rod using the non-Fourier damped wave conduction and relaxation equation.

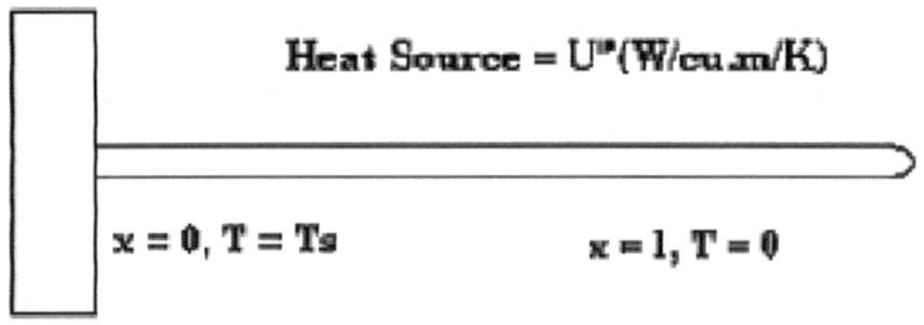

Figure 1.0 Temperature Dependent Heat Source in a Rod

A temperature dependent heat source with the source strength U''' w/m^3/K is present in the rod. The energy balance on a thin section with thickness Δx is considered at a distance x from the origin for an incremental time Δt. Thus in 1 dimension,

$$(q A|_x - q A|_{x+\Delta x} + U''' T A \Delta x) \Delta t = A \Delta x (\rho C_p) \Delta T \qquad (20)$$

Dividing throughout the equation wrt x and t and taking the limits as Δx, Δt goes to zero and at constant cross-sectional area, the energy balance equation becomes;

$$-\partial q/\partial x + U'''T = ((\rho C_p) \partial T/\partial t \qquad (21)$$

The non-Fourier damped wave heat conduction and relaxation equation can be written as;

$$q = -k\partial T/\partial x - \tau_r \partial q/\partial t \qquad (22)$$

The governing equation for the temperature can be obtained by eliminating the heat flux between the energy balance equation and the constitutive law for heat conduction. Thus differentiating the energy balance equation wrt to time and the constitutive equation wrt to x and eliminating the second cross-derivative of flux wrt to time and space;

$$U'''T + k\partial^2 T/\partial x^2 = ((\rho C_p \tau_r) \partial^2 T/\partial t^2 + (\rho C_p - U'''\tau_r)\partial T/\partial t \qquad (23)$$

Using the dimensionless variables;

$$u = T/T_s ; \qquad \tau = t/\tau_r; \; X = x/\mathrm{sqrt}(\alpha\tau_r) \qquad (24)$$

The governing equation in temperature becomes;

$$U'''u + S\partial^2 u/\partial X^2 = S\partial^2 u/\partial\tau^2 + (S - U''')\partial u/\partial\tau \qquad (25)$$

Let U* = U'''/S, Then the dimensionless governing equation can be written as;

$$U^*u + \partial^2 u/\partial X^2 = \partial^2 u/\partial\tau^2 + (1 - U^*)\partial u/\partial\tau \qquad (26)$$

Where, $S = (\rho C_p/\tau_r)$ is the storage coefficient. It has units of w/m^3/K. It can be seen from the governing equation that when S = U''', the damped wave conduction and relaxation equation simplifies to a wave equation. The equation reverts to the governing equation seen for the finite slab at constant wall temperature when U''' = 0. Thus when U'''/S = 1,

$$U^*u + \partial^2u/\partial X^2 = \partial^2u/\partial\tau^2 \quad (27)$$

Let $\eta = X^2 - \tau^2$

For $X > \tau$, the governing equation will transform to ;

$$U^*u + 4\eta\, \partial^2u/\partial\eta^2 + 4\partial u/\partial\eta = 0 \quad (28)$$

$$\eta^2\, \partial^2u/\partial\eta^2 + \eta\partial u/\partial\eta + U^*\eta u/4 = 0 \quad (29)$$

Comparing the above equation with the generalized Bessel equation; a = 1; b = 0; s = ½; d = U*/4; c = 0. p = 2. sqrt(0) = 0 ; sqrt(d)/s = $U^{*1/2}$.

$$u = c_1 J_0 (U^{*1/2}(X^2 - \tau^2)^{1/2} + c_2 Y_0 (U^{*1/2}(X^2 - \tau^2)^{1/2} \quad (30)$$

c_2 can be seen to be zero as u is finite at zero η.

Thus, $$u = c_1 J_0 (U^{*1/2}(X^2 - \tau^2)^{1/2} \quad (31)$$

This function exhibits damped wave behavior. This is valid till the first zero. Thus at the first zero of the Bessel function,

$$5.7831 = U^*(X^2 - \tau^2)$$

Thus for, $$X \geq \text{sqrt}(5.7831/U^* + \tau^2) \quad (32)$$

The temperature u will be zero. For short times there will be a good portion of the rod that will not have any temperature. This is even at infinite heat source strength. This is a clear manifestation of the finite speed propagation of the heat. For $\tau > X$,

$$u = c_1 I_0 (U^{*1/2}(\tau^2 - X^2)^{1/2} \quad (33)$$

From the boundary condition at X = 0,

$$1 = c_1 I_0 (U^{*1/2}(\tau) \quad (34)$$

Eliminating c_1 between the two equations an approximate solution for u can be written as;

$$u = I_0 (U^{*1/2}(\tau^2 - X^2)^{1/2}/ I_0 (U^{*1/2}(\tau) \quad (35)$$

The general solution for the temperature in the rod with a temperature dependent heat source can be obtained as follows. Let the solution be expressed as a sum of steady state and transient state components of the dimensionless temperature.

Let, $$u = u^{ss} + u^{\tau} \quad (36)$$

Then Eq. [25] can be written as;

$$U^*u^{ss} + \partial^2 u^{ss}/\partial X^2 = \partial^2 u^{\tau}/\partial \tau^2 + (1 - U^*)\partial u^{\tau}/\partial \tau - U^* u^{\tau} + \partial^2 u^{\tau}/\partial X^2 \quad (37)$$

The steady state component will obey the equation;

$U^*u^{ss} + \partial^2 u^{ss}/\partial X^2 = 0$ with the boundary conditions; (38)

$$X = 0, \; u^{ss} = 1 \quad (39)$$

$$X = X_l \; u^{ss} = 0 \quad (40)$$

The solution to the second order ODE will then be;

$$u^{ss} = c'\mathrm{Sin}(U^{*1/2}X) + c''\mathrm{Cos}(U^{*1/2}X) \quad (41)$$

From the boundary condition given in Eq. [39], c" can be seen to be 1. From the boundary condition given in Eq. [40],

$$c' = -\mathrm{Cot}(U^{*1/2}X_l) \quad (42)$$

Thus the steady state solution to the temperature is given by;

$$u^{ss} = \mathrm{Cos}(U^{*1/2}X) - \mathrm{Cot}(U^{*1/2}X_l)\,\mathrm{Sin}(U^{*1/2}X) \quad (43)$$

$$\text{or } u^{ss} = \mathrm{Cos}(U^{*1/2}X_l)\,(\,\mathrm{Cos}(U^{*1/2}X)/\,\mathrm{Cos}(U^{*1/2}X_l) - \mathrm{Sin}(U^{*1/2}X)/\,\mathrm{Sin}(U^{*1/2}X_l)) \quad (44)$$

III B. Critical Point of Zero Transfer

It can be seen that at steady state the temperature is periodic with respect to position. This is an interesting result. It can also be noted that the mathematical expression given in Eq. [44] can take on negative values. Negative temperature cannot exist as according to the third law of thermodynamics the lowest temperature attainable is 0 K. At 0 K the entropy of any system would be zero. The interpretation of the solution given in Eq. [44] is that after a certain location in the rod the temperature will be zero. This can be referred to as the critical point of zero heat transfer. This is shown in Figure 2.0. This was generated using a Microsoft excel for windows software package on a 1.9 GHz pentium IV personal computer. In the Figure 2.6, the heat source $U^* = U'''/S$ was taken as 0.5, length of the rod 10 cm, thermal diffusivity 10^{-5} m^2/s, relaxation time, τ_r as 15 seconds. For $X \geq 3.75$ the temperature comes to the end temperature, of) K imposed on the right end of the rod. Beyond this region there is no heat transfer. It can also be noted that the temperature within the rod exceeds the surface temperature. In this case, this is caused by the temperature dependant heat source. The maximum in temperature occurs at, X = 1.5. Further since the temperature is higher within the rod, the heat flux direction will be in the reverse direction of what it was to begin with neat the zero time. Thus the heat flow will be from the maximum location at X = 1.5 toward the X = 0 location as well as from the X = 1.5 location to the $X = X_l$ location. To begin with the problem was one where the initial temperature was at 0 K. The surface temperature at X = 0, was T_s. At short times the heat flux would be from the X = 0 location toward the right hand side of the surface. This has been reversed by the time the system reaches steady state. The heat source term is the contributor to the energy. The heat flux expression at steady state can be written as;

$$q^* = -\partial u^{ss}/\partial X = U^{*1/2}Cos(U^{*1/2}X_l)\ (\ Sin(U^{*1/2}X)/\ Cos(U^{*1/2}X_l) \quad +\ Cos(U^{*1/2}X)/ Sin(U^{*1/2}X_l)) \tag{45}$$

where, $q^* = q/sqrt(k\rho C_p/\tau_r)/(\ T_s)$

It can be seen from Figure 3.0 that several things happen when the steady state heat flux is plotted as a function of the distance in the rod , according to the solution given in Eq. [2.394]. The following distinctions can be recognized from this illustration; There exists a location of maximum heat flux.

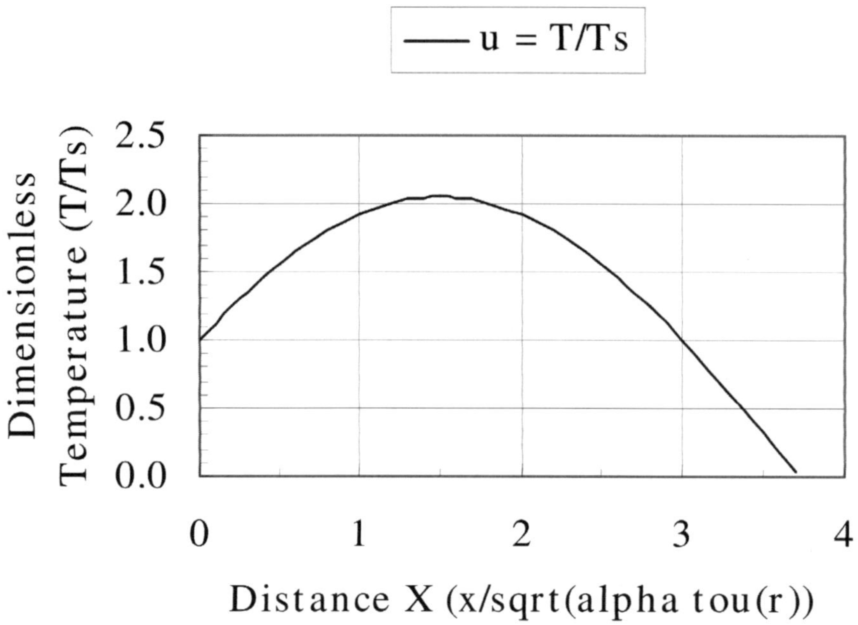

Figure 2.0 Dimensionless Temperature along a Heated Rod

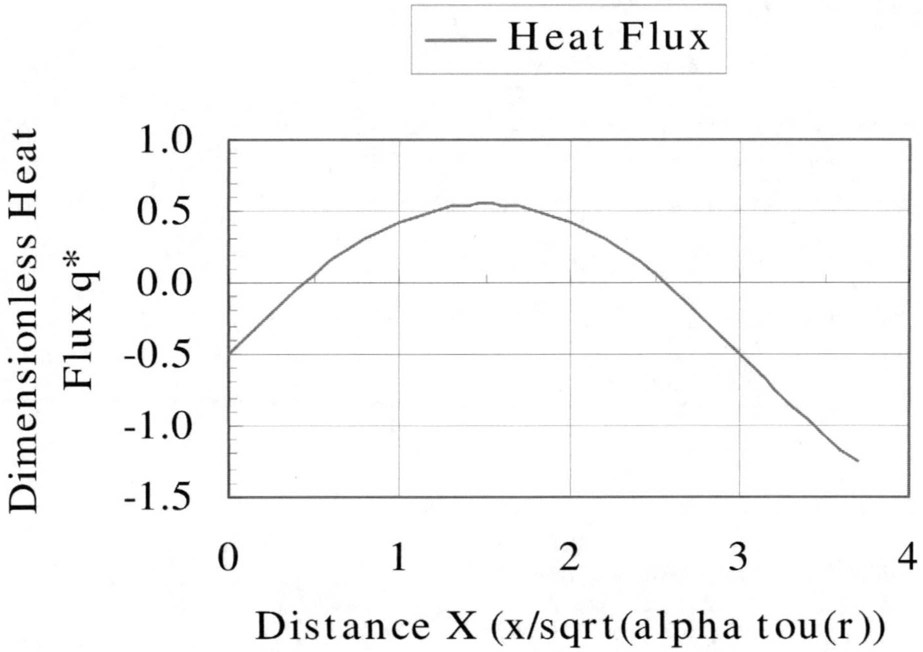

Figure 3.0 Dimensionless Heat Flux along a Heated Rod

In Figure 3.0, for the assumed values of the length of the rod, the relaxation time, the ratio of heat source strength with the storage coefficient, the location where the maximum heat flux occurs is at the dimensionless distance, X = 1.5.

There is a found a critical location in the rod beyond which there is no heat transfer. This is found to occur at X = 3.75 in Figure 2.0, Figure 3.0. There is found two locations in the rod where the heat flux changes in direction. The cross-over locations occur at X = 0.5 and again at X = 2.55. These are locations of minimum heat flux. The transient portion of the solution will then be;

$$\partial^2 u^\tau/\partial\tau^2 + (1 - U^*)\partial u^\tau/\partial\tau = U^* u^\tau + \partial^2 u^\tau/\partial X^2 \tag{46}$$

The boundary conditions are;

$$X = 0,\ u^\tau = 0 \tag{47}$$

$$X = X_l,\ u^\tau = 0 \tag{48}$$

Eq. [46] can be solved for by the method of separation of variables. First the damping term is removed by the substitution, u^{τ} = wexp(-τ/2). Eq. [25] becomes,

$$(1 - U^*)(\partial w/\partial \tau - nw) + (\partial^2 w/\partial \tau^2 + n^2 w - 2n\partial w/\partial \tau) = U^* w + \partial^2 w/\partial X^2 \quad (49)$$

$$\partial^2 w/\partial \tau^2 + \partial w/\partial \tau\,(1 - U^* - 2n) + w(-n(1 - U^*) + n^2 - U^*) = \partial^2 w/\partial X^2 \quad (50)$$

Letting $n = (1- U^*)/2$, Eq. [50] becomes;

$$\partial^2 w/\partial \tau^2 - w/4(1+ U^*)^2 = \partial^2 w/\partial X^2 \quad (51)$$

Eq. [51] can be solved by the method of separation of variables.

$$\text{Let } w = V(\tau)\phi(X) \quad (52)$$

Eq. [51] becomes;

$$V''/V - 1/4(1+ U^*)^2 = \phi''/\phi = -\lambda_n^2 \quad (53)$$

The solution for the second order ODEs can be written as follows;

$$\phi = c_1 \mathrm{Sin}(\lambda_n X) + c_2 \mathrm{Cos}(\lambda_n X) \quad (54)$$

From the boundary condition given in Eq. [47], $c_2 = 0$. From the boundary condition given in Eq. [48],

$$c_1 \mathrm{Sin}(\lambda_n X_l) = 0 \quad (55)$$

$$(\lambda_n l/\mathrm{sqrt}(\alpha \tau_r)) = n\pi, \quad n = 0,1,2\ldots\ldots \quad (56)$$

$$\lambda_n = n\pi\, \mathrm{sqrt}(\alpha\tau_r)/l \quad (57)$$

The time portion of the solution can be written as;

$$V = c_1 \exp\tau/2\mathrm{sqrt}((1+ U^*)^2 - 4\lambda_n^2) + c_2 \exp{-\tau}/2\mathrm{sqrt}((1+ U^*)^2 - 4\lambda_n^2) \quad (58)$$

The transient portion of the solution will decay out to leave the steady state portion of the solution alone. The zero temperature at the x = l, does the job of removing heat as it is generated in the rod. Thus at infinite time w = u^{τ} exp(τ/2) = 0 times infinity = 0. Thus w = Vϕ at infinite time = 0. Therefore V = 0 at steady state. Hence the constant c_1 is zero in Eq. [58].

$$\text{Thus, } V = c_2 \exp{-\tau}/2((1+ U^*)^2 - 4\lambda_n^2) \quad (59)$$

The general solution can be written as;

$$u^{\tau} = \Sigma_0^{\infty} c_n \exp(-\tau(1 - U^*)/2)\exp(-\tau/2((1+ U^*)^2 - 4\lambda_n^2)\mathrm{Sin}(\lambda_n X) \quad (60)$$

where, λ_n is given by Eq. [57]. The c_n can be solved for from the initial condition and is found to be;

$$c_n = 2[1 - (-1)^n]/n\pi \qquad (61)$$

Thus the general solution can be written as;

$$u = \mathrm{Cos}(U^{*1/2}X_l)\,(\,\mathrm{Cos}(U^{*1/2}X)/\,\mathrm{Cos}(U^{*1/2}X_l) \;-\; \mathrm{Sin}(U^{*1/2}X)/\,\mathrm{Sin}(U^{*1/2}X_l))$$

$$+\; \Sigma_0^{\infty}\,[1 - (-1)^n]\; 2/n\pi \exp(-\tau(1-U^*)/2)\, \exp(-\tau/2((1+U^*)^2 - 4\lambda_n^{\,2})\mathrm{Sin}(\lambda_n X)$$

$$(62)$$

where, λ_n is given by Eq. [57]. It can be seen from the general solution that even for n = 1 in the infinite series, when;

$$\lambda_n > (1+U^*)/2 \quad \text{or when} \quad 1 < 2\pi \mathrm{sqrt}(\alpha\tau_r)/(1 + U^*) \qquad (63)$$

the temperature will undergo subcritical damped oscillations. This is for the cases when $U^* < 1$. It can be seen that for $U^* > 1$, the "damping term" will begin to grow in amplitude with time and become a runaway.

Conclusions

The damped wave conduction and relaxation equation can be derived from the Stokes Einstein formulation for dilute solutions and slowly moving sphere by analogy between mass diffusion and heat conduction. The velocity of mass that was derived from accounting for the acceleration of the sphere in a fluid medium was found to be identical with the root mean square velocity of molecules of a ideal gas from the kinetic representation of pressure. A formula for the relaxation time of mass was derived in terms of diffusion coefficient, molecular weight, temperature and molar gas constant. For low pressure systems the relaxation time can be expected to be high.

The violation of laws of thermodynamics that has been reported by other investigators when mathematical model solutions were attempted to be derived from the non-Fourier hyperbolic PDEs were earlier shown to be a feature of the inconsist time conditions used to solve for the integration constants. When a steady state wave temperature condition is used well bounded solutions well within the second law of thermodynamics was found.

In this study a temperature dependent heat source is considered in a fuel rod. Even at steady state the temperature at some locations becomes negative violating the third law of thermodynamics. Only the Fourier's law was used. Evidently there is a zone beyond which there will be a null transfer of heat.

A temperature dependent heat source is imposed on a rod held at O K on one end and constant temperature on the other. A Bessel composite function solution results and the critical length above which the temperature will remain unchanged is derived when $U^* = 1$. This is when, $X \geq \mathrm{sqrt}(5.7831/U^* + \tau^2)$. A steady state temperature profile is derived. The heat flux is obtained for the rod heated by a temperature dependent heat source. A heat flux reversal was found after a said distance in the rod. A critical point of zero transfer was found. The method of separation of variables was used to obtain the general solution for the temperature profile. A bifurcated solution is found. For length smaller than, $l < 2\pi \mathrm{sqrt}(\alpha\tau_r)/(1 + U^*)$, subcritical oscillations in temperature was found.

References

[1] Sharma, K. R., "Manifestation of Acceleration During Transient Heat Conduction", *Journal of Thermophysics and Heat Transfer,* Vol 20, n4, 2006, 799-808.

[2] Maxwell, J. C., "On the Dynamical Theory of Gases", *Philosophical Transactions of the Royal Society of London, B: Biological Sciences,* Vol 157, 1867, 49-.

[3] Cattaneo, C., "A Form of Heat Conduction which Eliminates the Paradox of Instantaneous Propagation", *Comptes Redndus Hebdomadaires des Seances de L'academiedes Sciences*, Vol 247, 1958, 431- 433.

[4] Vernotte, P., "Les Paradoxes de la Theorie Continue de l'Equation de la Chaleur", *Comptes Redndus Hebdomadaires des Seances de L'academiedes Sciences*, Vol 246, n 22, 1958, 3154- 3155

[5] Sharma, K. R., *Damped Wave Transport and Relaxation*, Elsevier, Amsterdam, 2005.

[6] Ali, A. H., "Statistical Mechanical Derivation of Cattaneo's Heat Flux Law", *Journal of Thermophysics and Heat Transfer*, Vol 13, n 4, 1999, 544-545.

[7] Ali, A. H., "Non-Fourier Heat Flux Law in Diatomic Gases", *Journal of Thermophysics and Heat Transfer*, Vol 14., n2, 2000, 281-283.

[8] Glass, D. E. and McRae D. S., "Variable Specific Heat and Thermal Relaxation Parameter in Hyperbolic Heat Conduction Equation" Vol 4, n2, *Journal of Thermophysics and Heat Transfer,* 1990, 252-254.

[9] Nernst W, *Die Theoretician Grundalgen desn Warmestages* Knappe Hall, Frankfurt, 1917.

[10] Ozisik, M. N., Tzou, D. Y., "On the Wave Theory of Heat Conduction", ASME Winter Annual Meeting, Anaheim, CA, 1992.

[11] Bai, C., and Lavine, A. S., "On Hyperbolic Heat Conduction and the Second Law of Thermodynamics", *AMSE Journal of Heat Transfer*, Vol. 117, n2, 1995, 256-263.

[12] Landau, L., and LifshitzE. M., *Fluid Mechanics*, Pergamon, Oxford, UK, 1987.

[13] Tzou, D. Y., *Macro to Microscale Heat Transfer: The Lagging Behavior*, Taylor and Francis, Washington, DC, 1997.

[14] Kamisnki, W., "Hyperbolic Heat Conduction Equation for Materials with a non-Homogeneous Inner Structure", *ASME Journal of Heat Transfer*, Vol. 112, n3, 1990, 555-560.

[15] Mitra, K., Kumar, S., Vedavarz , A., and Moellemi, M. K., "Experimental Evidence of Hyperbolic Heat Conduction in Processed Meat", *ASME Journal of Heat Transfer*, Vol. 117, n2, 1992, 719-726.

[16] Taitel, Y., "On the Parabolic, Hyperbolic and Discrete Formulation of of the Heat Conduction Equation", *International Journal of Heat and Mass Transfer*, Vol. 15, n2, 1972, 126-128.

[17] Barletta, A., and Zanchini, E., "Hyperbolic Heat Conduction and Local Equilibrium: A Second Law Analysis", *International Journal of Heat and Mass Transfer*, Vol 40, n5, 1997, 1007-1016.

[18] Jou, D., Casas-Vasquez, J., Lebon, G., "Extended Irreversible Thermodynamics", *Report of the Progress in Physics*, Vol. 51, 1988, 1105-1179.

[19] Sharma, K. R., *Temperature Dependent Heat Source Effect in Cooling Nuclear Reactors,* 225th ACS National Meeting, New Orleans, LA, 2003, American Chemical Society, Washington, DC..

[20] Sharma, K. R., *On the Derivation of an Expression for Relaxation Time from Stokes-Einstein Relation*, 233rd ACS National Meeting, Chicago, IL, March 2007, American Chemical Society, Washington, DC.

[21] Stokes, G. G., *Mathematical and Physical Papers,* Cambridge University Press, UK, 1880.

[22] Einstein, A., *Brownian Movement*, Dover, New York, 1956.

[23] Serway, R. A., Faughn, J. S., Vuille, C., Bennet, C. A., *College Physics*, Brooks Cole, New York, 2005.

[24] Maxwell, J. C., Garber, E., Brush, S. G., Everitt, C. W. F., *Maxwell on Molecules and Gases*, MIT Press, Boston, MA, 1986.

[25] Halliday, D., Resnick, R., and Walker, J., *Fundamentals of Physics*, Wiley, New York, 2004.

230a Experimental and Analytical Investigation of Heat Transfer Phenomena in the Chain Section of a Rotary Lime Kiln

Masahiro Kawaji[1], Samer Hassan[1], Michael Massad[1], and Honghi N. Tran[2]. (1) Department of Chemical Engineering and Applied Chemistry, University of Toronto, 200 College Street, Toronto, ON M5S 3E5, Canada, (2) Pulp & Paper Centre, Dept. of Chemical Engineering & Applied Chemistry, University of Toronto, 200 College Street, Toronto, ON M5S 3E5, Canada

In a typical lime kiln used in pulp and paper mills, steel chains are attached to the inner wall of the lime kiln to enhance heat transfer from a hot combustion gas to a wet lime mud bed which move countercurrent to each other. The mud enters at about 30 oC with some mositure, and undergoes drying, heating, and calcination reaction steps, while the hot gas enters at ~ 1000 oC and leaves the kiln at about 200 oC.

This work is aimed at increasing the performance of rotary lime kilns by accelerating the drying process of a wet lime mud using a chain system in order to initiate the calcination reaction more quickly and increase the lime production rate. To increase the mud drying rate, curtains of steel chains are currently installed in the drying section to be heated by the hot gas and release the stored heat when the chains move through the wet mud as shown in Fig. 2. A systematic investigation was initiated to understand the thermal interactions occurring between the chain and hot gas, and between the chain and the wet mud.

The objective of the current work is to develop a calculation scheme to estimate the total heat transfer rate for a given chain system consisting of thousands of chain rings based on a lumped heat capacity technique. Following a review of relevant literature (Brimacombe and Watkinson, 1978; Gorog et al., 1981; Boateng and Barr, 1996; Jamialahmadi and Malayeri, 1995; Palmer and Howes, 1998), a thermal model was developed to predict the temperature history of a single chain ring at a certain position in a given chain and the amount of energy transferred as the chain is alternately exposed to the hot gas and wet mud. The total heat transfer enhancement for an entire chain system could then be estimated as the sum of the heat transfer rates for all chain rings.

Lumped Heat Capacity Model

A lumped heat capacity model was used to analyze the effects of the ring diameter and its position in a chain on the heat transfer rates from the hot gas to the wet mud as illustrated in Fig. 3. The temperature variations of the chain ring while in the gas and wet mud were calculated for different residence times and heat transfer coefficients of the chain ring in the hot gas and wet mud of varying moisture contents. From the lumped heat capacity model predictions, the temperature of a chain ring was found to increase with increasing chain diameter, gas velocity and gas temperature. Furthermore, for various positions of a single ring within a chain, the exposure time to gas would drastically change. For instance, a chain ring located at the very top of the chain would receive the maximum exposure time, since its residence time within the lime mud would be minimal. On the other hand, the ring at the bottom of the chain would remain in the mud much longer and be exposed to the hot gas only when the chain is freely suspended from the top of the kiln. The residence times of chain rings in the gas and mud are thus critical parameters in the prediction of the overall heat transfer rate for the entire chain system along with the effective heat transfer coefficients in the gas and mud.

In the lumped heat capacity model, an unsteady energy equation was solved numerically to predict the variation of the chain ring temperature with time. The maximum and minimum chain temperatures were found to be sensitive to the exposure time in the gas and heat transfer coefficient in the mud, respectively. The difference between the maximum temperature in the gas and minimum temperature in the mud determines the amount of heat transferred from the hot gas to the mud by the chain ring. The actual exposure time in the gas and residence time in the mud would depend not only on the location of the chain ring but also clustering of nearby chains, since there are a total of 32 chains attached to the inner periphery of the lime kiln in a plane at the hot end and 40 chains in a plane at the cold end of the chain section. The clustering phenomenon could effectively shield a given chain ring from the hot gas and shorten the exposure time.

Laboratory Experiment

In order to assess the effective exposure time in the presence of neighboring chains, a mock up of the chain system in one plane has been constructed. The chains used were 3/8-inch chains made from a brass rod with a diameter of 0.8 mm. A bed of sand was placed inside the cylinder to simulate the lime mud.

As the cylinder was rotated on a pair of rollers, there appeared to be a significant amount of mud caught on each ring, thus reducing the heat transfer rate from the hot gas to the chain ring. Also, due to the tumbling effect the number of fully exposed chains on the wayward side was reduced. Furthermore, the chain strands that were not freely hanging could be observed to cluster near the kiln wall. At any given time during a single rotation, the rings at the very bottom of any chain strand were not exposed. The total number of chains used in this system was 40, however, during one snapshot it was found that only 8 received the optimum exposure time.

Further experiments were conducted to gain insights into the interactions among the neighboring chains and with the mud in order to best estimate the effective exposure time of a given ring in the gas and residence time in the mud. Three out of 20 pairs of chain strands were labeled and the exposure times of the top and bottom halves of each chain strand in the gas and mud were measured during one rotation (60 seconds). The average exposure time to gas was roughly 21 seconds for the top half of the chain strand, while it was only 11 seconds for the bottom half. On the other hand, the average residence time in the mud was 19 seconds for the bottom half of the chain strands, slightly longer than 16 seconds for the top half.

When a chain was not exposed to the gas or mud, it was clustered amongst neighboring chains. The clustering durations were 16 and 19 seconds for the top and bottom halves, respectively, and slightly shorter than the exposure time in the gas for the top half but greater for the top half. Thus, clustering with neighboring chains could significantly reduce the exposure times in the gas and mud for both the top and bottom halves of the chain. This implies that the effectiveness of the chains in transferring heat from the hot gas to the mud could be severely limited if the chain density is high.

REFERENCES

Boateng, A. A. and Barr, P.V., "A thermal model for the rotary kiln including heat transfer within the bed", Int. J. Heat Mass Trans. 39 (10), 2131-2147 (1996).

Brimacombe J. K.and Watkinson, A. P., "Heat Transfer in a Direct-Fired Rotary Kiln: I. Pilot Plant and Experimentation", Metallurgical Transactions B, 9B, 201-219 (1978).

Gorog, J. P., Brimacombe, J. K., Adams, N. T., "Radiative heat transfer in Rotary Kilns", Metallurgical Transactions B, 12B, 55-69 (1981).

Jamialahmadi, M., and Malayeri, M. R., " Prediction of Heat Transfer to Liquid-Solid Fluidized Beds", Can. J. Chem. Eng., 73, 445-455 (1995).

Palmer, G. and Howes, T., "Heat Transfer in Rotary Kilns", Cement Industry Federation Technical Conference (1998).

230c Enhanced Heat Transport in a Pulsating Heat Pipe Using a Self-Rewetting Working Fluid

Koji Fumoto, Mechanical Engineering, Kushiro National College of Technology, Otanoike W2-32-1, Kushiro, 084-0916, Japan and **Masahiro Kawaji**, Chemical Engineering and Applied Chemistry, University of Toronto, 200 College Street, Toronto, ON M5S 3E5, Canada.

New experimental results will be reported on the enhancement of heat transport by a pulsating heat pipe (PHP) using a self-rewetting fluid as a working fluid. Self-rewetting fluids have a property that the surface tension increases with temperature unlike other common liquids. The increasing surface tension at a higher temperature means that the liquid will be drawn towards a heated surface if a dry spot appears. Thus, in boiling, a dryout phenomenon may be prevented at a higher heat flux. In the present experiment, butanol was added to water at a concentration of less than 1 wt% to make the self-rewetting fluid. A pulsating heat pipe made from an extruded multiport tubing was partially filled with the butanol-water mixture and tested for its heat transport capability at different input power levels. One end of the PHP was heated using a copper block equipped with two cartridge heaters. A fin was attached to the opposite end to be cooled by a fan. The experiments showed that the maximum heat transport capability was enhanced by a factor of four when the maximum heater temperature was limited to 120 degrees C. Thus, the use of a self-rewetting fluid in a PHP has been shown to be highly effective in improving the heat transport capability of pulsating heat pipes.

230d Characteristics of Freezing Heat Transfer of Pcms in Elliptical Cylinders

Subrahmaniyam Susarla, Chemical Engineering, OSMANIA UNIVERSITY,UNIVERSITY COLLEGE OF TECHNOLOGY, 2/2/647/155A,CENTRAL EXCISE COLONY,, NEW NALLAKUNTA,, Hyderabad-500007, India

Freezing of single materials is an important thermal process in several equipment and in thermal energy storage systems.Freezing studies pertaining to cylinders under various thermal conditions are numerous in literature.The importance of study of freezing in non-circular crosssectional cylinders was noted.We wish to present a comparative study based on experiments in cylinders of elliptical and circular crosssections. Two PHASE CHANGE MATERIALS(PCM) namely n-octadecane(MP 28C) and

paraffin wax (MP 58-60C)are being used.The range of variables to be covered are: volume of PCM:50cc to 1000cc,L/d:1 to 5,temperature drop across the PCM:2 to 20C,initial superheat of PCM:0 to 25C. Elliptical cylinders of eccentricity variation 0.4 to 0.95 are fabricated in copper.Experimental data would include frozen mass variation with time and temperature distribution(in two dimensions)in the PCM during freezing.Earlier work of the author indicated accurate correlations of data of frozen mass/area with the product of (melting point-wall temperature)xtime by power law type expressions for n-Octadecane.However the proportionality constant appears to be a complex function of the physical properties of the pcm.The data of paraffin wax show significant deviatios from the correlation and the theoretical equations.Both frozen mass variation with time and temperature distribution with time for circular and elliptical cylinders would be discussed.

230e Enhanced Heat Transfer in Vertical Thermosiphon Reboiler

Omid Zadakbar, Ali Vatani, and Kianoosh Karimpour. Faculty of Chemical Engineering, University of Tehran, Enghelab St., Tehran, Iran

ABSTRACT

Vertical thermosiphon reboilers play the important role in heat exchange processes and are widely used in various processes. Hydrostatic pressure on the fluid at entrance to tubes causes forming of subcooled zone in vertical thermosiphon reboiler tubes. Since heat transfer coefficients in subcooled zone comparing to boiling heat transfer coefficients are significantly low, the presence of subcooled zone causes the overall heat transfer coefficient in reboiler tubes to be reduced. Appling enhanced heat transfer increase overall heat transfer coefficient inside the tubes of vertical thermosiphon reboilers and significantly eliminates performance problems. In fact heat transfer enhancement decrease the subcooled zone length and surface temperature of tubes which reduce fouling formation in this zone.

In our previous works, we have studied on the effect of using several types of tube inserts in a subcooled zone of a thermosiphon reboiler and in this paper we study on applying some different tube inserts in subcooled zone of a vertical thermosiphon reboiler. Having prepared a simulated model, we studied on the effect of using wire matrix insert, helically coiled Inserts and twisted tape inserts in this model. Finally, effect of these tube inserts on overall heat transfer coefficient, subcooled zone overall heat transfer coefficient, average heat flux rate, maximum temperature of reboiler tube, subcooled zone length and heat transfer surface area in a vertical thermosiphon reboiler have been discussed.

ارامكو السعودية
Saudi Aramco

Energy Saving in Crude Oil Pre-heat Train for Predetermined Retrofit-Based Target

Abdullah El-Mesfer
*Ahmed S. Aseeri
GAP-TECH
Dammam
KSA
Dr. Mahmoud Bahy Noureldin
Consulting Services Department
Saudi Aramco
Dhahran
KSA

AIChE Spring Meeting
New Orleans,

April 6-11, 2008

Acknowledgment

Deep Gratitude to Consulting Services department of Saudi Aramco

Colleague Dr. Mahmoud Bahy Noureldin, Energy Systems unit/ Consulting services department

Colleague Ahmed Saleh Aseeri of GAP-TECH

ارامكو السعودية
Saudi Aramco

Outline

- Introduction
- Refinery Pre-Heat Train
- Problem Statement
- Pinch Technology Approach
- Mathematical Programming Approach
- Combined Constraint Satisfaction and Repeated Simulation Approach
- Result and Discussions
- Conclusion

ارامكو السعودية
Saudi Aramco

Crude Distillation Unit Overview

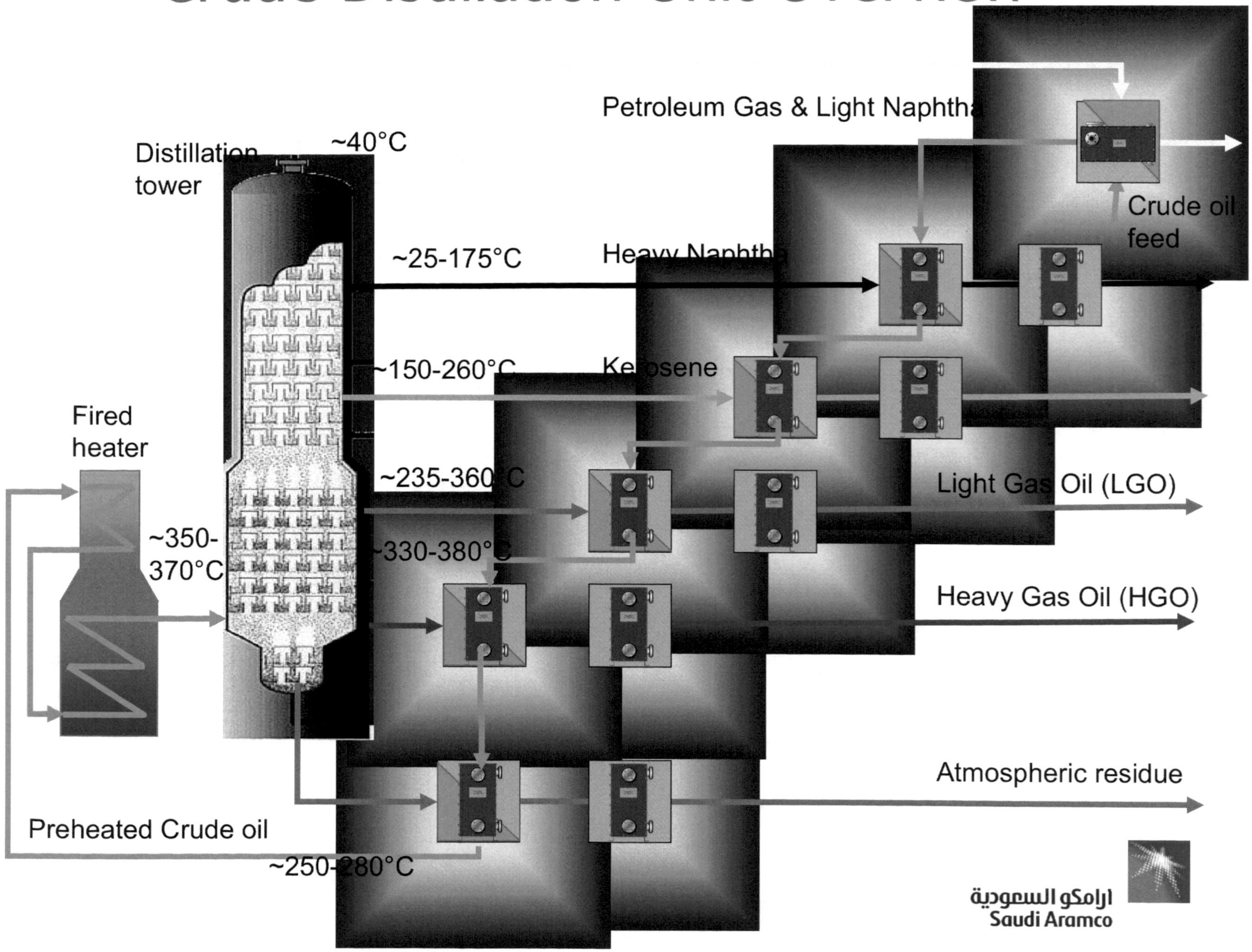

ارامكو السعودية
Saudi Aramco

Crude Pre-Heat Train

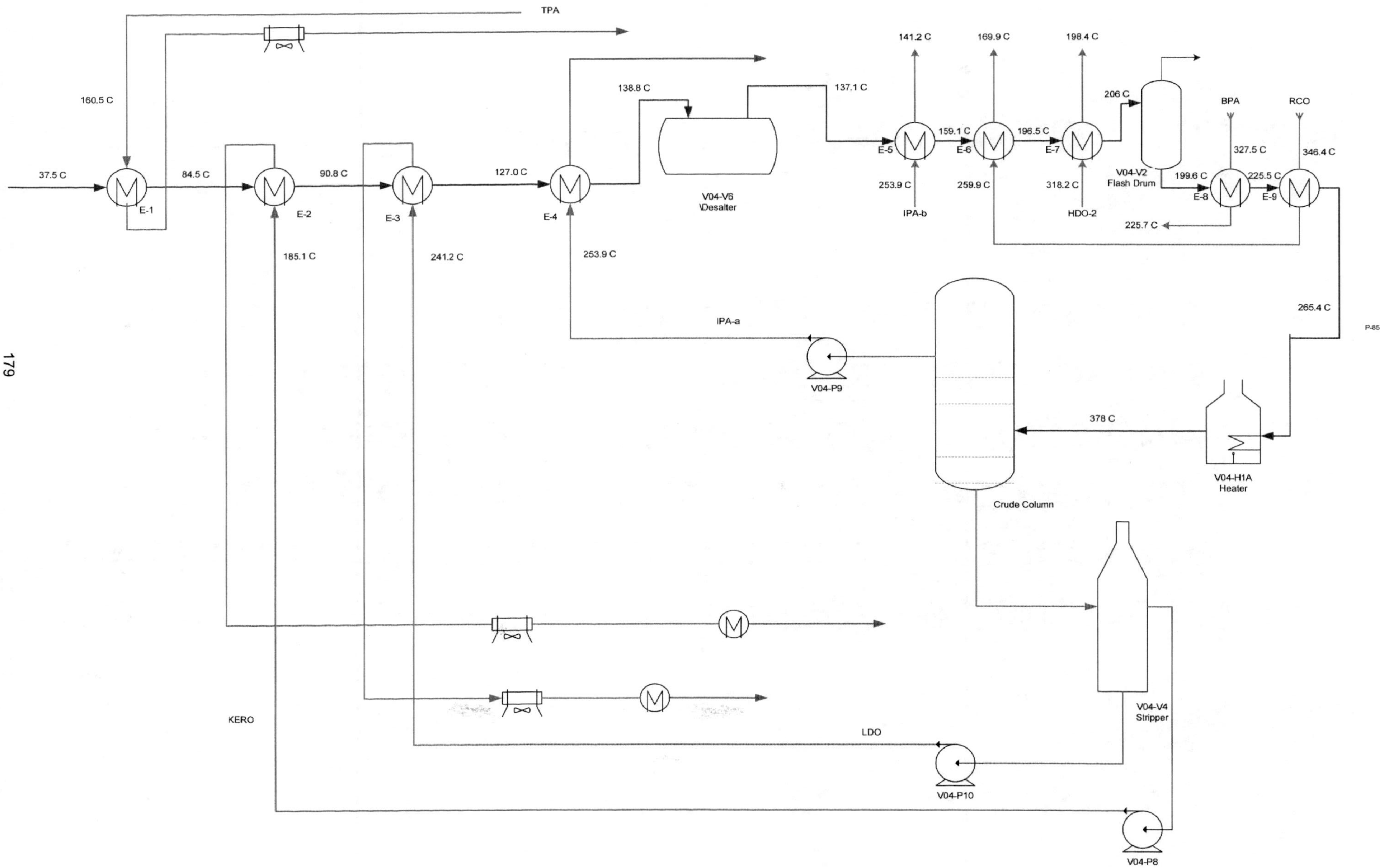

Problem Statement

"Given a crude pre-heat train of an existing facility, retrofit with minimum possible modifications the HEN in a way that maximizes energy recovery in the pre-heat train and minimizes GHG emissions. The retrofit actions taken should preserve plant's maximum safety and reliability as well as productivity needs.

A maximum of two splits in the crude stream is allowed"

ارامكو السعودية
Saudi Aramco

Crude Pre-Heat Train

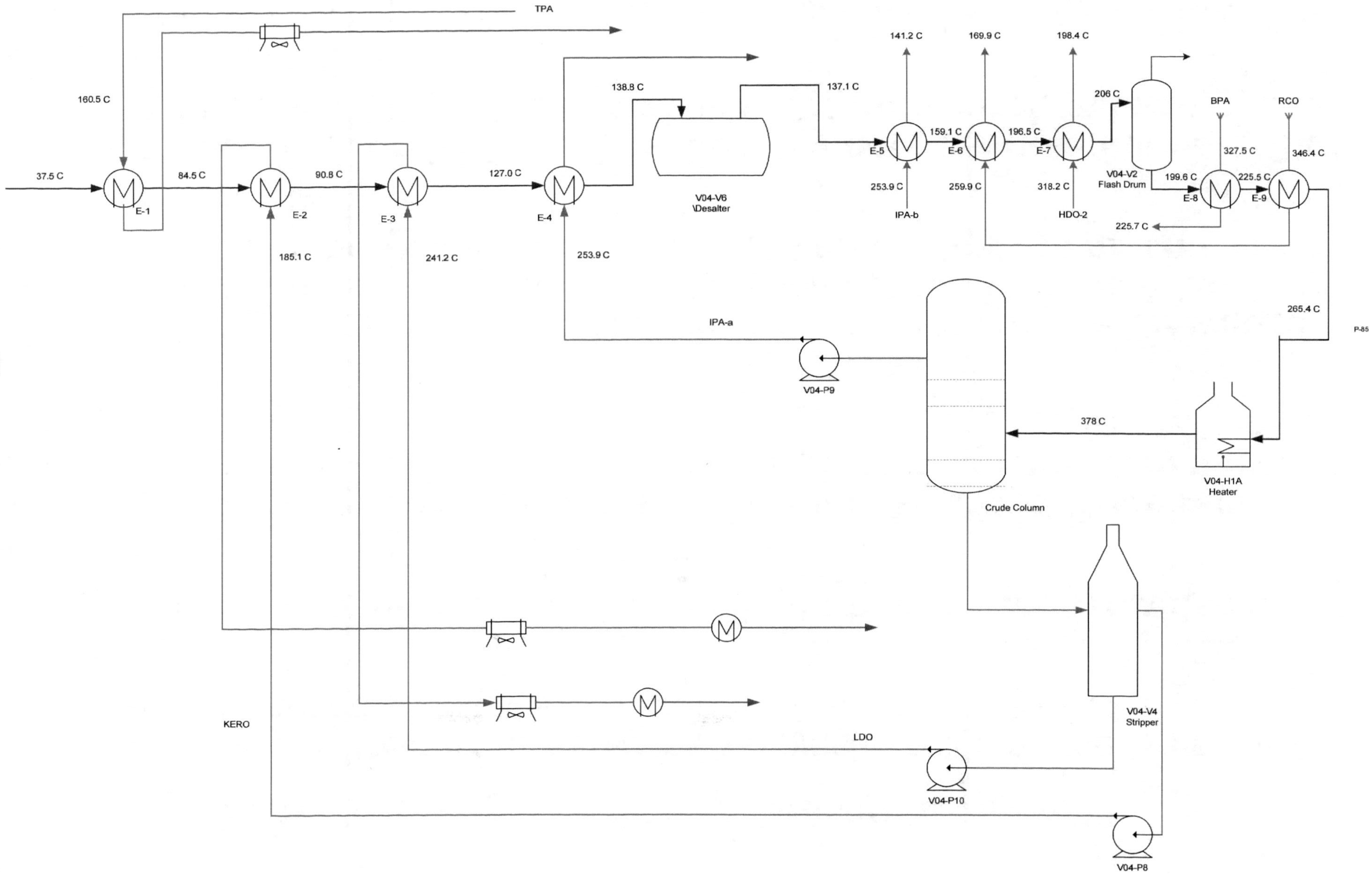

Pinch Technology Method

This design procedure is known as the pinch design method and can be summarized as follows:

Divide the problem at the pinch into two separate problems

The design for the separate problems is started at the region of minimum choice known as the pinch point and then moves away

Temperature feasibility requires constraints on the CP values to be satisfied for matches between streams "at the pinch" for the two problems above the pinch and below the pinch

The loads on individual heat exchangers are determined using the tick-off heuristic to minimize the number of units

Away from the pinch there is usually more freedom in the choice of matches. In this case the designer can choose on the basis of his/her process knowledge

Grassroots Solution as Retrofit Approach

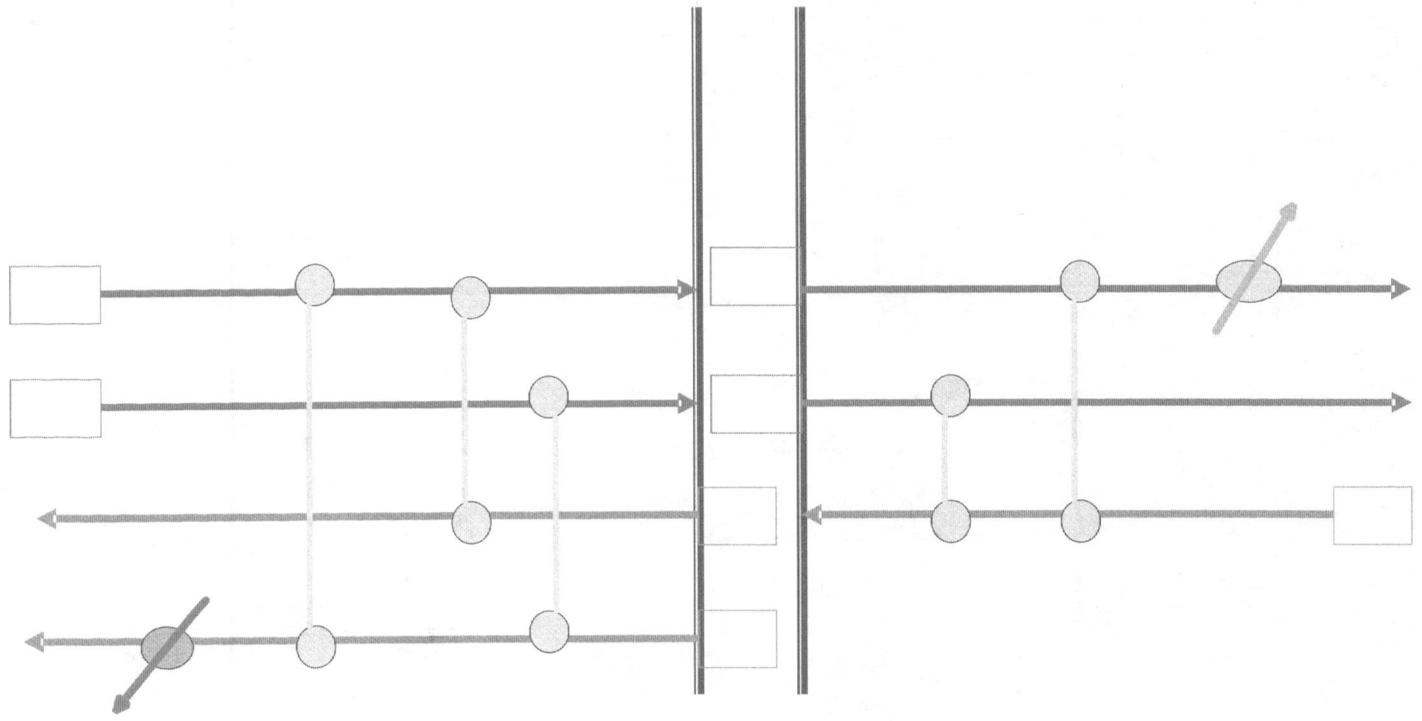

Scheme I

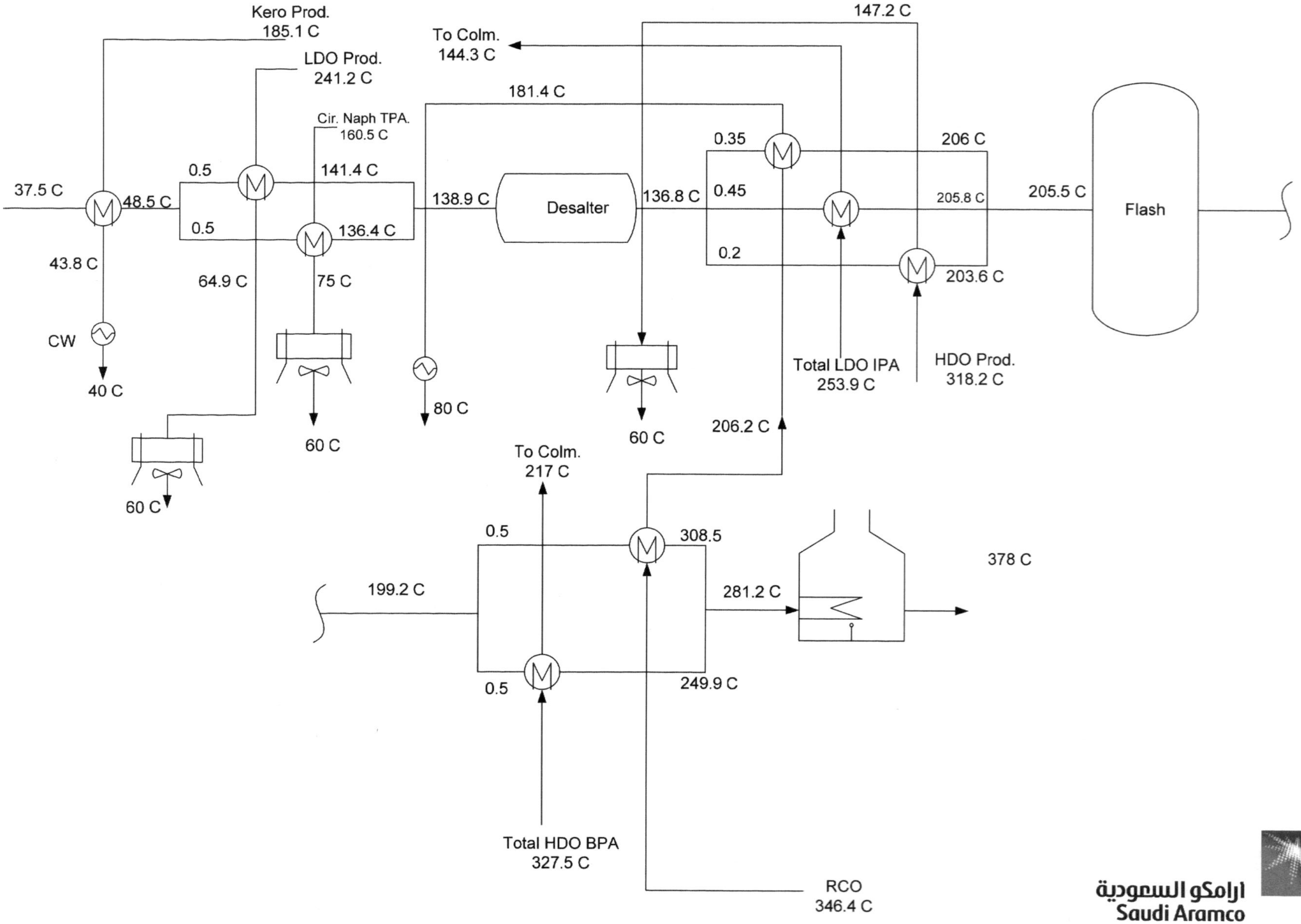

ارامكو السعودية
Saudi Aramco

Scheme II

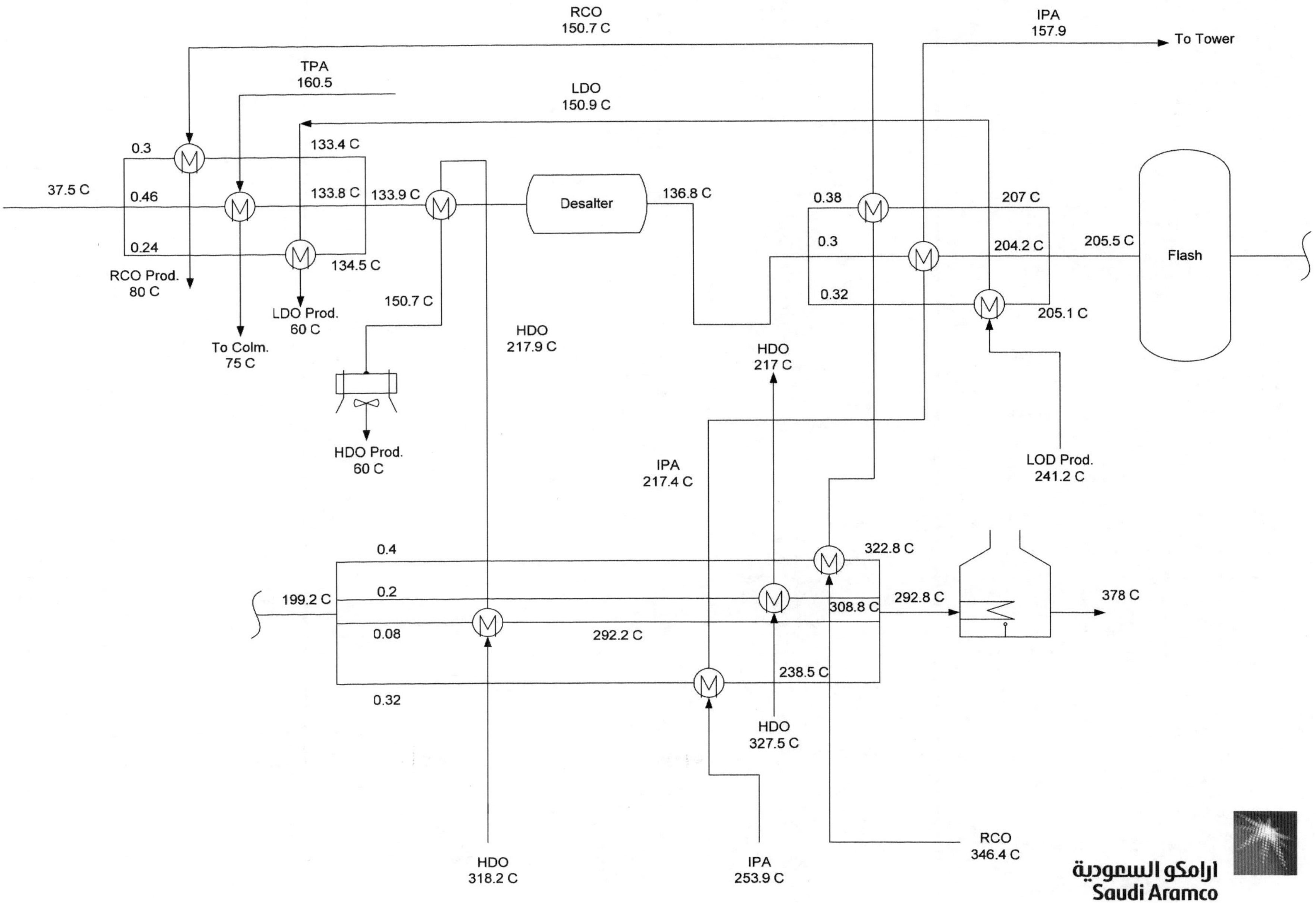

ارامكو السعودية
Saudi Aramco

Scheme III

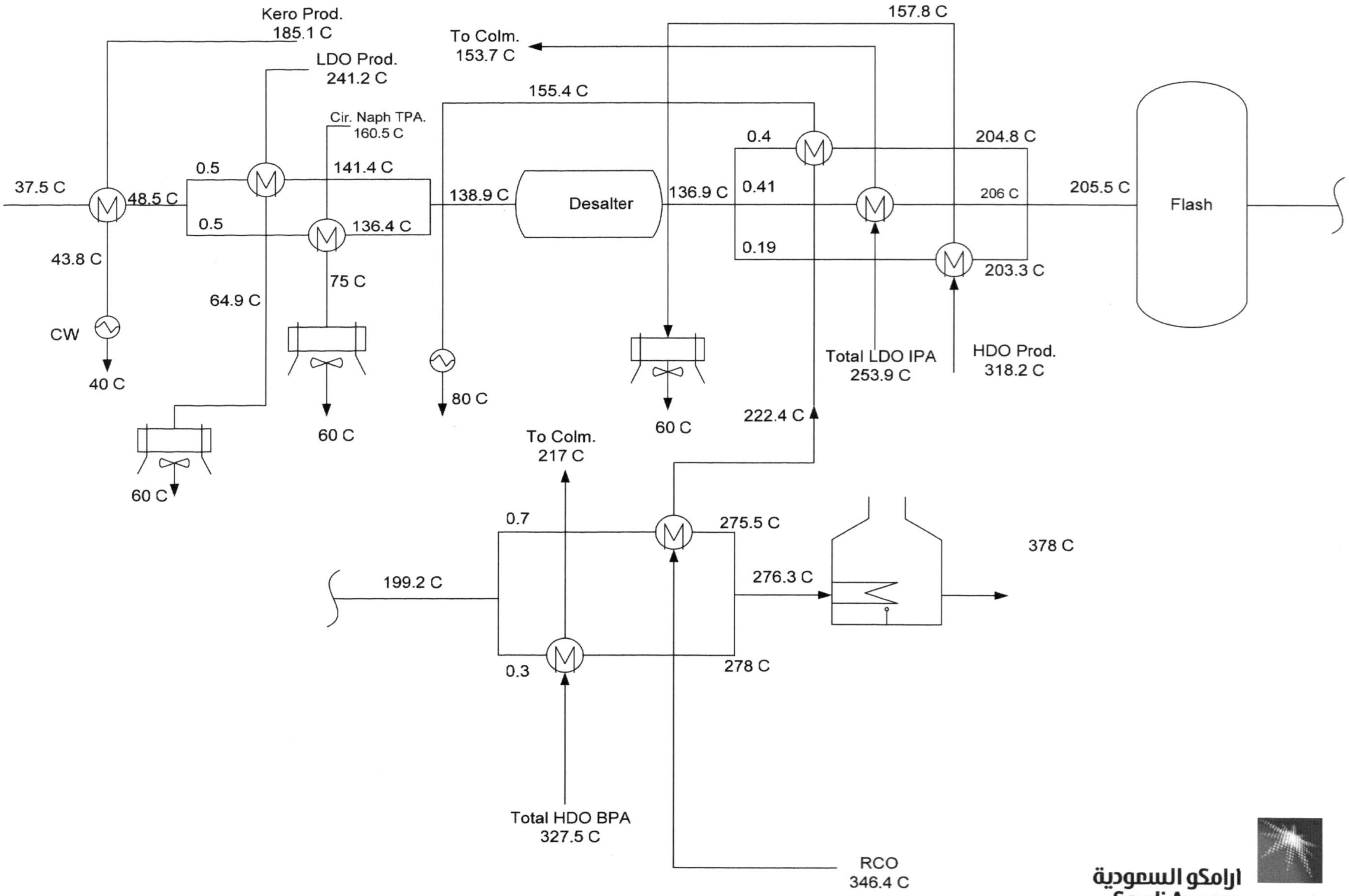

Kero Prod.
185.1 C
LDO Prod.
241.2 C
Cir. Naph TPA.
160.5 C
To Colm.
153.7 C
155.4 C
157.8 C
37.5 C
48.5 C
0.5
0.5
141.4 C
136.4 C
43.8 C
64.9 C
75 C
CW
40 C
60 C
60 C
138.9 C
80 C
Desalter
136.9 C
60 C
0.4
0.41
0.19
204.8 C
206 C
203.3 C
205.5 C
Flash
Total LDO IPA
253.9 C
HDO Prod.
318.2 C
222.4 C
To Colm.
217 C
0.7
0.3
199.2 C
275.5 C
276.3 C
278 C
378 C
Total HDO BPA
327.5 C
RCO
346.4 C

Scheme IV

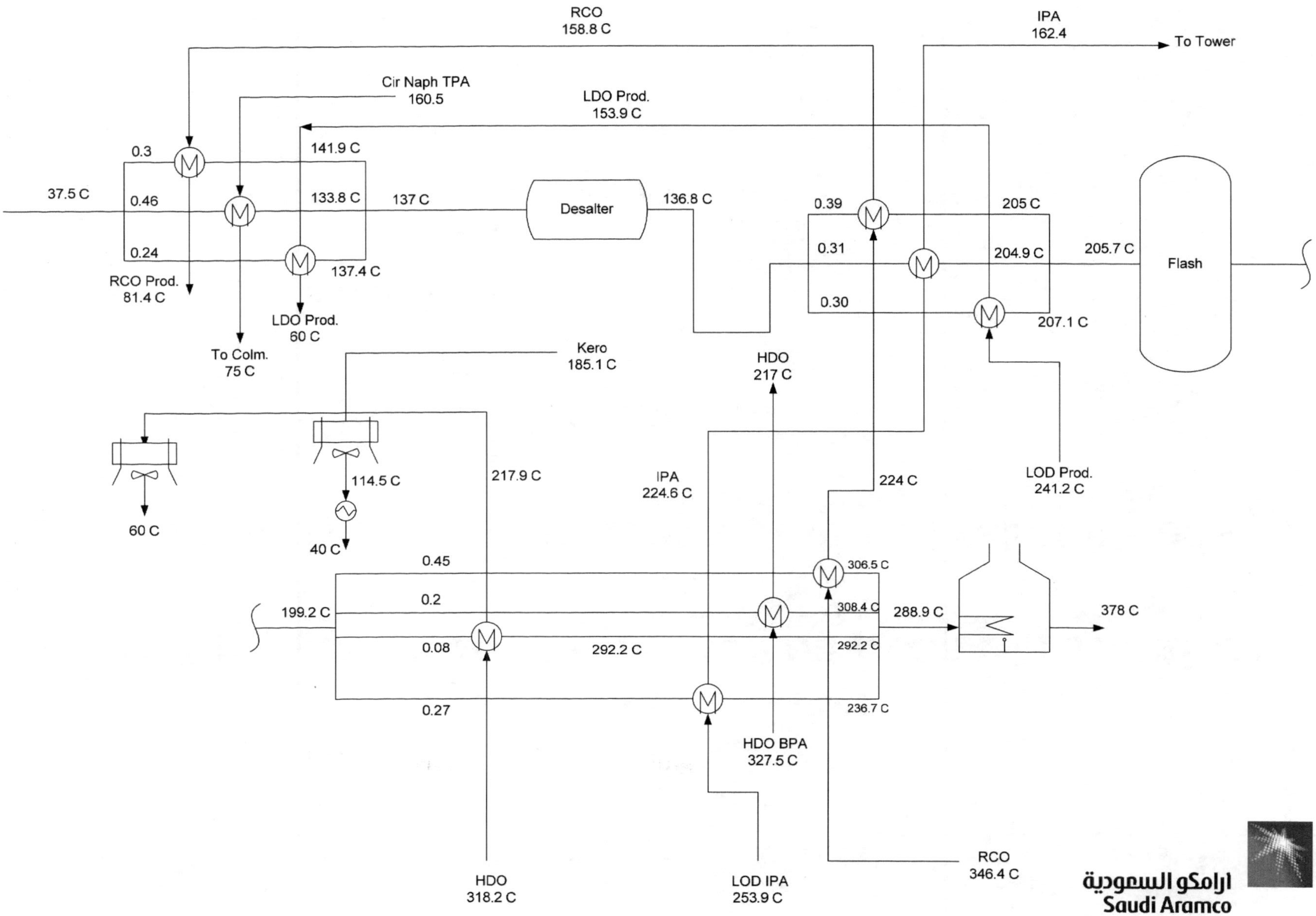

ارامكو السعودية
Saudi Aramco

Mathematical Programming Approach

Superstructure that embed all possible solutions

Objective Function that include all relevant costs

And a Mathematical model

Minimize cost

s.t.

Model equations and constraints

HEN Superstructure for Crude Stream

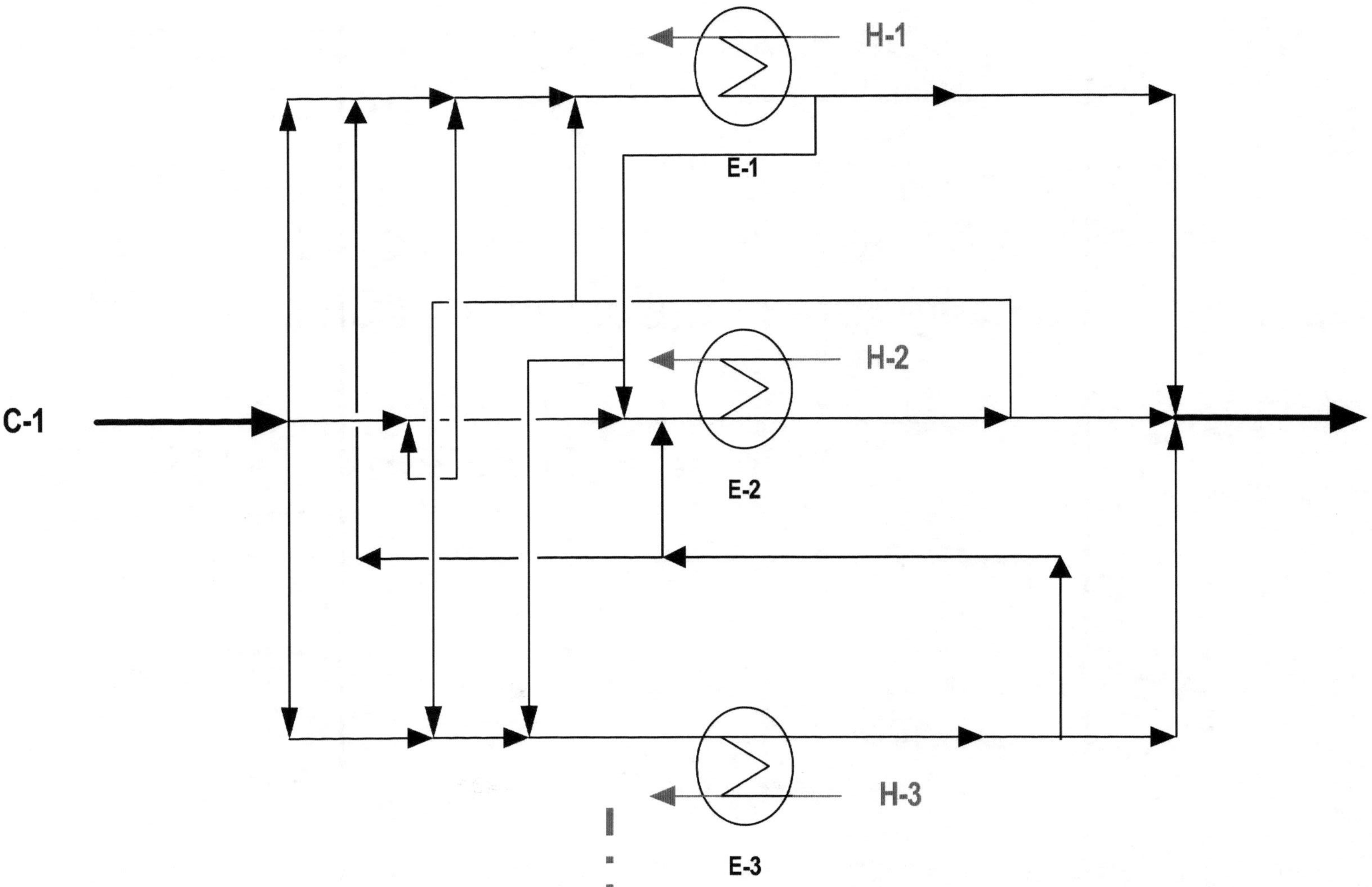

Saudi Aramco

Combined Constraint Satisfaction and Repeated Simulation Approach

The essence of this method is as follows:

•CS is used first to eliminate all non-feasible configurations from the superstructure and provide tighter lower and upper bounds on the continuous variables

then

•RS to techno-economically evaluates feasible configurations and define the optimal configuration

Note: Second step can be conducted by MP, however the constraints imposed by The process owners made the repeated simulation technique easier

Pre-Heat Train Temperature Interval Diagram

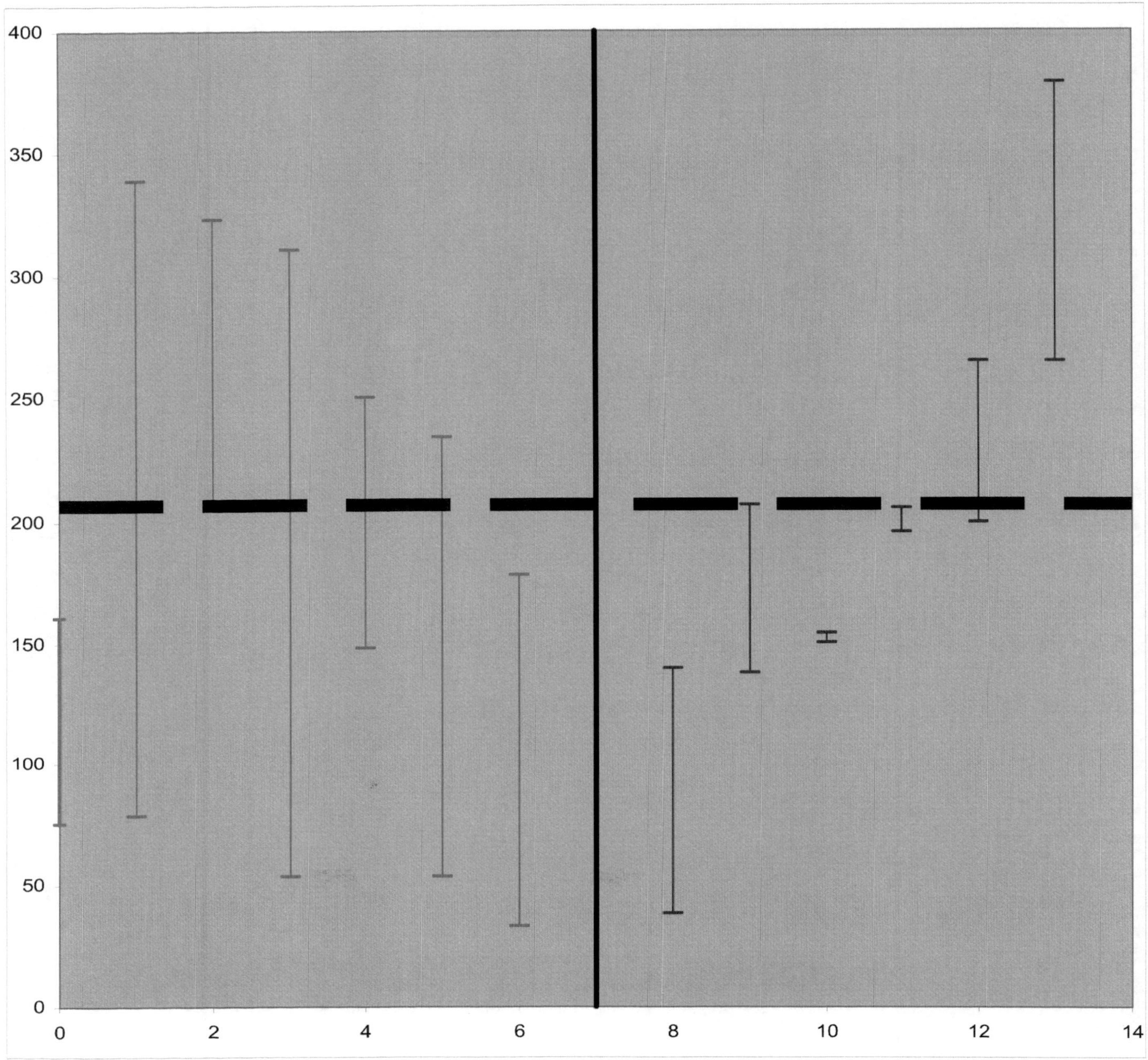

Result and Discussion

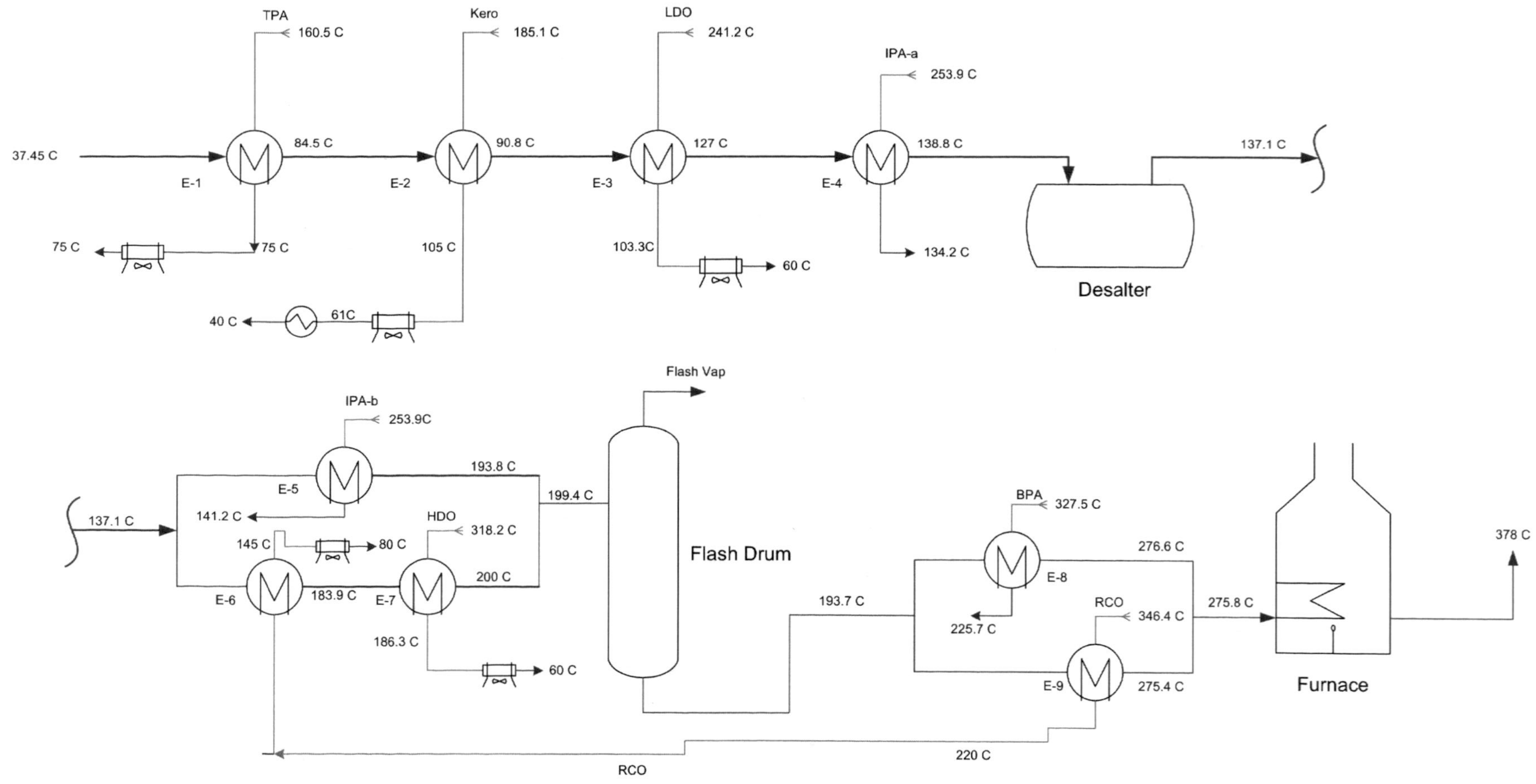

ارامكو السعودية
Saudi Aramco

Results and Discussion (continued)

- Only two splits of crude stream in the possible three locations
- One split per location
- Higher temperature to crude heater by 10.4°C
- Energy saving of 14 MMBtu/hr
- Carbon dioxide reduction By 10000 tonnes/yr

ارامكو السعودية
Saudi Aramco

Conclusion

- A little modification to old refineries can bring in appreciable levels of energy saving and carbon dioxide emission reduction
- Refined products supplies are tight and this trend will continue for years to come
- Retrofit projects for energy savings, while vital, are very difficult to get approved by refineries
- Process owners in many situations dictate acceptable levels of retrofit

ارامكو السعودية
Saudi Aramco

Standards for Automated Demand Response Communications

David G. Holmberg
Stephen J. Treado

Abstract

While electric utilities have been offering a wide variety of programs to curtail load in times of stress on the electric grid, there have been no standards for program parameters, nor standard methods for communicating between utility and customer. In the past year there has been an effort to bring the utility providers and facilities industry together to reach agreement on communication technology and message content to allow for standardized communications. The Demand Response Automation Server provides for automated demand response (DR), allowing a customer's facility energy management system to follow pre-programmed energy reduction measures in response to utility signals. These signals might provide notice of a load curtailment event or provide a real time price update. The benefits of this for industrial customers are the potential for participating in DR programs with multiple energy providers, reduced cost of energy management systems and hardware, and thus more options for implementing demand response in industrial facilities.

This paper introduces the Demand Response Automation Server and describes progress toward standardized communications. The role of ASHRAE's (American Society of Heating, Refrigeration and Air Conditioning Engineers) BACnet (Building Automation and Controls Network) communication standard in this effort, the tools within BACnet that serve demand response communications, and the place of BACnet relative to the industrial customer will also be discussed. Finally, some examples will be given how these standards can be used by the industrial customer to implement automated demand response.

Introduction

Energy consumption in industrial facilities is responsible for approximately 32 % of total energy consumption in the United States [1]. The relative contribution to CO_2 production is nearly proportional. Utility costs and the environmental impact of this energy consumption are becoming increasingly important to facility managers and owners. Most energy usage in industrial process control facilities is for process heat, and nearly 80 % of the electricity is consumed for process use (half of which powers motors and variable speed drives). However, 15 % of electrical energy use is for heating, ventilation and air-conditioning (HVAC) and lighting, while HVAC accounts for 7 % of total industrial energy usage [2]. HVAC and lighting are prime candidates for participation in demand response programs.

Demand Response (DR) activity has been defined as, "an action taken to reduce electricity demand in response to price, monetary incentives, or utility directives so as to maintain reliable electric service or avoid high electricity prices." [3] A utility customer can participate in DR in a variety of ways, from cutting non-essential loads and shifting loads via energy storage, to bringing on backup generation, and bidding that load to the electricity market.

The U.S. Department of Energy reports that 45 of the 50 states have demand response/ load management programs which provide incentives to curtail demand and reduce load during peak periods in response to system reliability (grid component failure) or market conditions (high demand and high price) [4]. On the federal level, congress just passed the Energy Independence and Security Act in December, 2007. Among the provisions of this bill is Title XIII "Smart Grid" which specifies a structure for coordinating standards and technology efforts around the country leading toward implementing the next generation electric grid. In the coming months, there will be increasing efforts at all levels of government and in private corporations to reach consensus on communication standards and implementation of the supporting technology.

Utility – Facility Communication Standards

Because industrial facilities represent such a large portion of a utility's electrical load, they can play an important role in promoting a more reliable electrical grid and an energy efficient future. This has important financial implications for both the facility owner and the utility provider. Considering that the average cost of electricity at peak can be ten times off-peak rates [5], there is a strong incentive for facility owners to cooperate in reducing peaks. Utilities benefit by reducing the need to build and operate expensive peaking power plants and by gaining more tools to use for improving the stability and reliability of the electric power supply. Effective, automated communication between utilities and facility energy management systems will be a critical component of realizing this potential.

Leaders in DR efforts include Lawrence Berkeley Laboratory's Demand Response Research Center (DRRC), OpenAMI (a user community developing the Advanced Metering Infrastructure for residential customers), investor owned utilities, and the Electric Power Research Institute. The DRRC has produced an initial draft standard interface for automated demand response for use in communicating with commercial customers. Their "Demand Response Automation Server" uses web services to send simple event shed information to demand response clients at customer facilities, which then act upon the shed request.

Demand Response Automation Server

The Demand Response Automation Server (DRAS) is a communication technology in development since 2002, and currently under test at the DRRC, with funding from the California Public Interest Energy Research (PIER) program. This past year DRRC began to work more closely with experts nationwide with the goal of establishing the DRAS as a standard for DR communications. DRAS is aimed at large electric customers (> 200 kW) in the commercial and industrial sectors. The DRAS is gaining prominence as a strong candidate for nationwide use as it is proving to be effective in reducing electrical consumption for industrial customers. Any facility with an Internet connection and an energy management control system (EMCS) is a candidate to make use of, and benefit from, the DRAS. The DRAS enables a large facility to participate in utility DR programs, as well as to bid power to the market. In either case the customer is able to sell power at a profit when electricity prices are high.

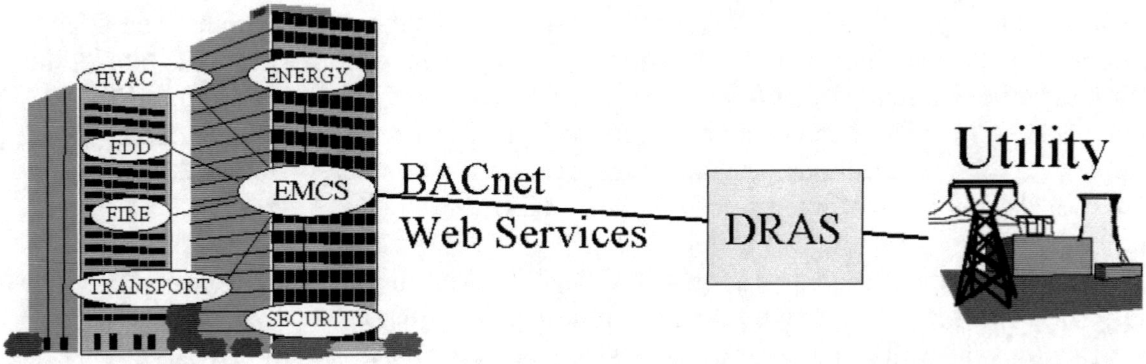

Figure 1 The DRAS is a standard interface between customer facility and utility information system.

The Demand Response Automation Server acts as a standard interface between utilities and customers to allow communications with any utility implementing that same interface (Figure 1). The standard covers three DR services: Critical Peak Pricing (CPP) event notifications, Real-Time Pricing (RTP), and Demand Bidding (DB). The standard gives the message details (in XML format) for each of these services, discusses the DRAS architecture and protocols for data transmission, as well as the roles of different actors on the utility and customer side.

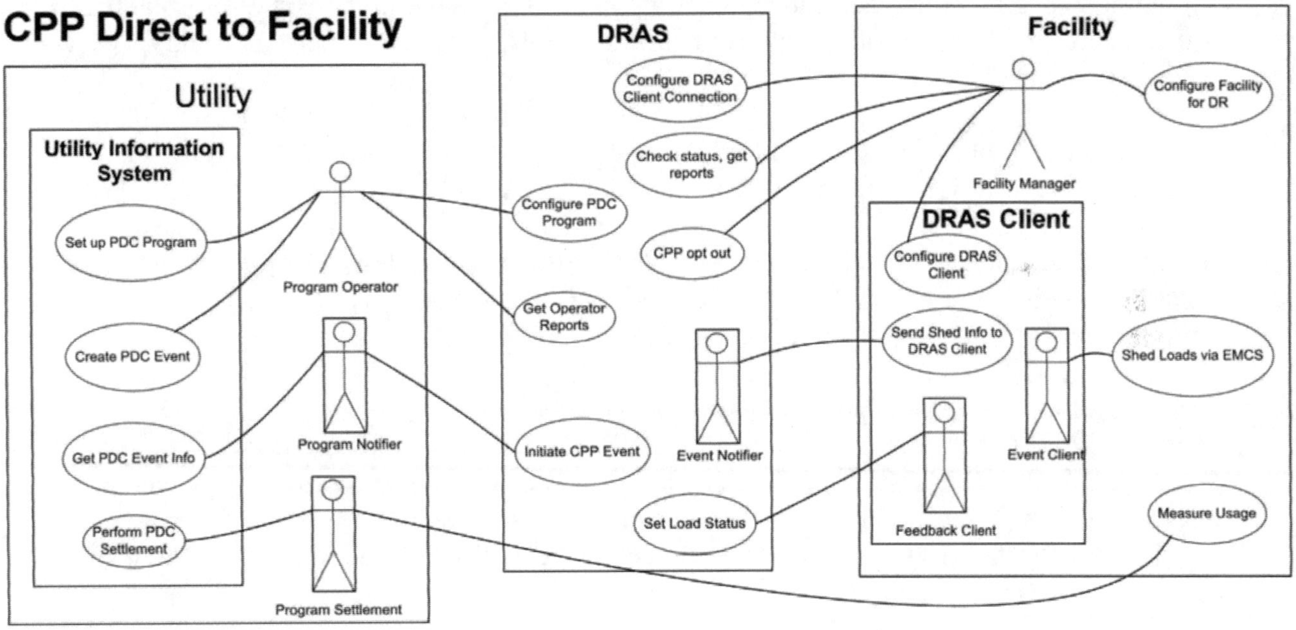

Figure 2 Critical Peak Pricing (CPP) example communication model for the DRAS. The DRAS acts as a standard interface between utility and customer to pass event notifications to customers. Machine actors are shown as boxed stick figures.

Figure 2 is the schematic from the DRAS draft [6] showing the actors and communication paths involved in the CPP event notifications. The three entities involved in the CPP event notification scenario are the utility on the left, DRAS interface, and customer facility on the right. The customer facility hosts the DRAS Client. The DRAS Client receives event notifications ("shed info") and gives feedback on shed status to the DRAS, while commanding automated shed actions on the facility side via the EMCS. The DRAS Client itself can be a physically separate box on the wall connected to the EMCS via relay connections, or via a wired network. Alternatively, the DRAS Client may be a software client integrated into the EMCS.

The DRRC has made the integration of the DRAS Client into the EMCS easier by selecting BACnet Web Services as the communication transport for shed event notifications (as well as all other messages between the DRAS and the DRAS client). The focus of the DRAS is on large commercial customers, which is also the focus of the BACnet standard. For this reason, the DRAS team has been working closely with the Utility Integration Working group of the BACnet standard committee.

BACnet for Facility Load Control

The Building Automation and Control Network (BACnet) ANSI/ASHRAE Standard 135 is an open and internationally recognized communication standard for building automation. BACnet began with a base in HVAC, and now also provides specific functionality for lighting, life safety, and access control, as well as specific tools for energy management. BACnet is an object-based protocol that handles building automation specific communications including: alarming, trending, notifications, reading values from sensors and writing values to control points, as well as other basic and advanced features.

One recent addition to the BACnet standard is the Load Control object [7]. The Load Control object is designed to provide a general, high-level interface used to manage shedding of electrical loads. The Load Control object's interface accepts commands that specify how much and when load reduction is required. Load Control objects can be linked in a distributed, hierarchical fashion for controlling arbitrarily complex combinations of electrical loads found throughout a facility.

The definition of the Load Control object does not specify how the electrical consumption is to be reduced or how consumption baselines should be determined. Whatever load reduction strategies are used, the EMCS will require configuration and/or programming in order to implement the chosen strategies. The Load Control object provides an interface to execute predetermined control actions and to view the current load shed status of a facility.

Consider an example of how a Load Control object could be used to reduce a facility's HVAC system electrical consumption. First, a mechanical engineer designs a sequence of operation for electrical demand reduction. That sequence is programmed into a supervisory controller in a facility control system, and a new Load Control object in that controller is configured to trigger the demand reduction sequence. An energy management application can write to properties of the Load Control object in a way that is equivalent to saying, for example, "reduce demand to 90 % of baseline for 180 min starting at 3 p.m." Processing this request will trigger a pre-programmed sequence of steps to reduce electrical load, and properties of the object will make visible the progress in complying with the requested load reduction.

It is expected that the Load Control object will become an important component of future automated demand response systems because it standardizes one communication interface between energy management systems and electrical loads.

The Role of BACnet Web Services

Web services have become a hot technology in the mainstream IT world because web services can make it easier to integrate computer applications over a network. An increasing number of popular business applications can function as web services clients, and modern software development tools make it easy to add web services functionality to custom software applications. Evolved from the World Wide Web Consortium (W3C), servers and clients using Web Services communicate via messages that are created with the Extensible Markup Language (XML) [8]; these messages are encapsulated using the Simple Object Access Protocol (SOAP) [9] for transmission over computer networks. Web services can play an important role in facilitating the flow of information between control systems and other systems.

The BACnet Web Services specification [10] was designed to make it easy for business applications to access data in building control systems. The data model was intentionally designed to be generic, allowing BACnet web services to be used with a wide variety of data sources, including control systems that do not communicate using BACnet. The data access services are also rather generic, allowing the web services client to read and write simple data values (integers, real numbers, strings, dates, etc.) and to retrieve trend data.

It is likely that BACnet web servers will become common components of building automation systems because they provide convenient access to building automation system data through an interface that is well understood by developers of a wide range of management software. As the details of the communication needs between facilities and utilities become clearer, it is expected that BACnet web services will be expanded and become the primary means for facility—utility interactions as part of a utility-industry standard.

Conclusion

Industrial sector energy consumption accounts for 32 % of total US energy use, and a significant percentage of that energy could be available for load shedding or load shifting in response to electricity price fluctuations. Energy costs, environmental impacts, and the need to manage electrical grids for reliability and stability are all strong drivers for automated electrical load management in buildings as well as automated communications with utilities. The BACnet standard supports load measurement and sophisticated load shedding strategies, while the Demand Response Automation Server promises to serve as a standard interface between energy management systems and utilities. In addition, the DRAS now implements BACnet Web Services to connect the DRAS to the customer facility.

Members of the BACnet Utility Integration Working Group are working with the Demand Response Research Center, and other national efforts to standardize communication between utilities and their customers. The end result will be an effective set of tools for utilities and facility managers to collaborate in reducing energy consumption and managing peak loads.

References

1. U.S. Energy Information Agency, *Energy Consumption by Sector Overview, 2005*, http://www.eia.doe.gov/emeu/aer/pdf/pages/sec2.pdf.

2. U.S. Energy Information Agency, *Manufacturing Energy Consumption Survey, 1991,* Table A36, ftp://ftp.eia.doe.gov/pub/consumption/industry/taba36a.pdf.

3. U.S. Federal Energy Regulatory Commission, Assessment of Demand Response and Advanced Metering 2007 Staff Report, September 2007.

4. U.S. Department of Energy, Office of Energy Efficiency and Renewable Energy, Federal Energy Management Program, Energy-Efficiency Funds and Demand Response Programs, http://www1.eere.energy.gov/femp/program/utility/utilityman_energymanage.html.

5. U.S. Government Accountability Office, *Electricity Markets: Customers Could Benefit from Demand programs, but Challenges Remain*, Aug. 2004, http://www.gao.gov/new.items/d04844.pdf.

6. California PIER Demand Response Research Center's Demand Response Automation Server, "DRAS Use Cases", July 2007, http://drrc.lbl.gov/dras/.

7. Addendum *e* to ANSI/ASHRAE Standard 135-2004. Available from http://www.bacnet.org/Addenda/index.html.

8. Extensible Markup Language (XML) 1.0 (Fourth Edition), available at http://www.w3.org/TR/xml/, September 2006.

9. SOAP Version 1.2 Part 1: Messaging Framework (Second Edition), available at http://www.w3.org/TR/soap12-part1/, April 2007.

10. Addendum *c* to ANSI/ASHRAE Standard 135-2004. Available from http://www.bacnet.org/Addenda/index.html.

Energy Systems Management and Greenhouse Gas Reduction

Stephen J. Treado
David Holmberg
National Institute of Standards and Technology

Abstract

The efficient use of energy provides the double benefits of controlling both energy costs and greenhouse gas (GHG) emissions. The industrial sector comprises about one third of United States GHG emissions, so presents an attractive opportunity for utilizing advanced design, operation and control strategies to minimize energy consumption while optimizing overall systems performance through the implementation of cogeneration, heat recovery, adaptive controls and other advanced techniques. One of the barriers to the utilization of these techniques is a lack of suitable design and analysis tools to guide their effective implementation. This paper presents an approach for optimizing overall systems design and operation using the principles of energy systems management to control energy usage and thereby reduce associated GHG emissions.

Introduction

Modern society is built on technologies that require useable energy input to provide a comfortable and safe environment, and to produce desirable goods and services. There are many different ways to identify, categorize and subdivide energy consuming sectors, either on the basis of end use, such as heating, cooling, lighting or manufacturing, or the general class of activity, such as transportation, industrial or buildings. However, regardless of how the energy consumption is categorized, the efficient use of energy is desirable. It is also true that energy use is not an end goal in itself, but is only the means to an end. The important point is that there are economic, environmental and social motivations for making efficient use of energy, while providing the desired outcome of the energy using activity. Focusing on energy efficiency as a means to control energy usage, rather than a simple reduction in energy consumption, allows us to meet our broader economic goals without sacrificing performance desired from the activity.

The link between energy efficiency and GHG emissions is clear. The principal GHG of concern is CO_2 which is a product of the combustion of carbon-based fuels [1]. Thus, both the efficiency of fuel burning devices and the efficiency of devices that use energy produced by other devices that burn fuel, affect GHG emissions. Taking a broader view, the efficiency of all the various processes in an energy conversion chain influences the total energy requirements, and therefore the associated GHG emissions.

Energy Management Systems

In most real world energy systems, there are multiple factors which can influence energy efficiency and performance. Some of these factors are somewhat static, such as combustion efficiency and heat transfer coefficients, while others are dynamic, such as thermal loads and environmental conditions. As a result, maintaining efficient system operation usually requires an energy management system (EMS) which can measure and manipulate essential system parameters [2]. Modern EMS's may also be capable of communicating with remote actors, such as utilities and weather forecasters, in order to perform sophisticated control actions, including load shedding, fuel switching, and other proactive measures. In order to make proper decisions and maintain optimum system operation, the EMS must be able to access all of the required information and have a means to evaluate alternatives and manipulate operating conditions. This implies a level of intelligence that is certainly feasible, but only recently seeing increased implementation.

This type of optimization goes beyond simple feedback control of individual processes, or even cascaded loop control, and ventures into intelligent hierarchical control strategies that can consider overall system performance, and adjust, activate or terminate processes as required [3]. Some of the EMS-based strategies to reduce energy usage and GHG emissions include:

- Controlling equipment and systems for higher efficiencies
- Reducing loads
- Shifting loads to off-peak time periods
- Energy storage
- Heat recovery
- Energy source selection

Methods such as cogeneration (combined heat and power), heat recovery and fuel switching can have a big impact on GHG emissions, and determining the best operating conditions requires careful consideration of many factors, including the carbon content of the fuel sources, the efficiencies and controllability of the various systems and components.

Energy Use and GHG Emissions

The total energy use by an enterprise may be composed of a number of components, but generally consists of a combination of electrical power obtained from the utility grid, combustion of fuel(s) on site, and possibly a heated or cooled fluid stream, such as steam or hot water, which is generated off site. In some cases, there may also be an on site source of electrical power from a renewable resource such as solar photovoltaic or wind power, or possibly a renewable heated fluid source, such as geothermal or solar thermal power. Another option is the on site generation of electrical power from a fuel source, such as a gas turbine or diesel engine, which may be accompanied by a byproduct source of heated fluid, a strategy called cogeneration. A schematic representation of offsite power generation with onsite heat source, compared to a cogeneration system, is given in Figure 1.

The GHG emissions that are associated with each of these energy sources can vary substantially, both compared to each other, and over time, as a function of the carbon content of the fuel. For example, electrical power generated by a coal-fired power plant would have a high CO_2 emission rate, while hydroelectric power would have essentially zero, while a natural gas fired generator would fall somewhere in between. The combined CO_2 emission rate would be a function of the particular mix of generating technologies in operation at that point in time.

When trying to evaluate the GHG emissions associated with the use of electrical power generated remotely, termed the source energy, it is necessary to know the effective carbon content of the mix of power generation sources. This is largely an accounting exercise, since power is fed into regional power grids from many sources, and extracted by multiple consumers. However, it is possible to obtain an estimate of the percentage contributions of each type of power generation, and from that determine the rate of CO_2 production as a function of electrical power consumed. However, these values can vary with time as different reserves are brought online to meet varying power demand. Similar values can be obtained for each primary energy source (i.e. each fuel source) for on site combustion. By comparing the CO_2 emission rates of the various energy sources, along with the conversion efficiencies, the desired mix of energy sources can be determined to minimize GHG emissions. As a side note, it would also be possible to minimize total energy usage, or energy cost, or any combination of the three parameters.

Heat Recovery, Energy Storage and Renewable Energy Sources

One of the advantages of on site power generation is the potential for combined heat and power systems (cogeneration). The large-scale electrical power version of this concept is known as district heating [4]. In either case, a fuel is burned to generate heat which powers a generator, and subsequently the waste heat is utilized for water, space or process heating. The combined efficiency of such a system is much greater than either power or heat generation alone. Depending on the circumstances, the use of cogeneration may result in lower GHG emissions, such as when the CO_2 emission rate of the electric utility is high, and there is a need for the waste heat from the power generation. This will frequently be the case for industrial installations, since process heating requirements can be very large. This would not be true, however, for industrial applications that had high electrical power requirements relative to heating requirements.

Energy storage is another concept that can be particularly beneficial. Whenever heating loads are variable and do not necessarily coincide with the availability of thermal energy, heat storage can increase overall system efficiency by allowing thermal energy to be stored and recovered for later use, thereby moving it from the waste to the useful category. Energy usage can also be shifted to times when the carbon content of the energy source is lower.

Renewable energy sources are always beneficial from a GHG reduction perspective, since they are assumed to have no net contribution to CO_2 production, and the energy they produce will substitute for energy sources that do emit GHG's. While higher energy conversion efficiencies are better since they provide greater output, the actual efficiency is not as important since it has an indirect effect on emissions that is related to the reduction in fuel use due to the substitution of renewable energy for non-renewable energy. Since renewable sources such as wind and solar radiation are essentially free, the conversion efficiency of renewable technologies is not as critical as the energy conversion efficiency of costly fuels.

Example- Cogeneration and GHG Emissions

One illustrative example of the potential for an EMS to reduce GHG emissions involves switching from offsite (electric utility) power source combined with onsite fuel based heating to onsite cogeneration (combined heat and power). The particular values used in the example are typical, but were selected to demonstrate the method rather than to evaluate the competing technologies. The metric for analysis is the rate of CO_2 production for an equivalent output of power and heat. The rate of CO_2 production ($\dot{m}_{CO2}$) is a function of the rate of the fuel used ($\dot{m}_F$) the carbon content (C, $\dot{m}_{CO2}/\dot{m}_F$), and the combustion (η_{comb}) and system (η_o) efficiencies. Every kilogram of carbon consumed produces 3.67 kg of CO_2. Thus:

$$\dot{m}_{CO2} = 3.67C\dot{m}_F / HV \tag{1}$$

Where *HV*- heating value of the fuel, kJ/kg

Since the electrical power delivered to the site is equal to the heating value of the fuel consumed times the system efficiency, which includes thermal, generation and transmission efficiencies, the rate of CO_2 production associated with a delivered electrical power of P_{EL} is given by:

$$\dot{m}_{CO2} = 3.67C \cdot P_{EL} / \eta_O HV \tag{2}$$

The same relation holds for the rate of onsite CO_2 production from combustion of fuel to produce heat, with the exception that system efficiency is replaced by combustion efficiency. This is usually a big difference numerically, since the system efficiency of electrical power is in the range of 25 to 40%, while combustion efficiencies are 70 to 90%.

The energy performance of cogeneration systems is characterized by the Energy Utilization Factor (EUF) [5], which is the ratio of the sum of the electrical power and rate of useful heat (Q_U) produced to the energy input from the fuel consumed (Q_F), and the Heat to Power Ratio (*HPR*), which is simply Q_U/P_{EL}. Using these relations, the rate of CO_2 production by the onsite cogeneration system is given by:

$$\dot{m}_{CO2,CG} = \frac{3.67C \cdot P_{EL}}{HV}\left(\frac{1 + HPR}{EUF}\right) \tag{3}$$

In a similar manner, the total CO_2 production rate for the offsite electrical power plus the onsite fuel-based heating is given by:

$$\dot{m}_{CO2,Off} = \frac{3.67C \cdot P_{EL}}{HV}\left(\frac{1}{\eta_O} + \frac{HPR}{\eta_{comb}}\right) \tag{4}$$

Equations 3 and 4 differ in the efficiency terms, and in the fact that in equation 3, HPR and EUF are linked (see below), but can be used to compare CO_2 generation for the two alternatives. A simple comparison is the ratio of $\dot{m}_{CO2,Off} / \dot{m}_{CO2,CG}$.

In order to focus on the system effects, as opposed to the fuel or component effects, the following assumptions were made:

- All fuel sources were natural gas, with HV=50MJ/kg, C=0.75, η_{comb}=0.80
- Electrical power requirement was constant at 1MW, but the heat requirement was varied
- The system efficiency for the offsite power generation was varied in three steps, 0.263, 0.35 and 0.40 (this is equivalent to reducing the carbon content of the fuel)
- The electrical generation efficiency for the cogeneration system was held constant at 0.263
- The baseline conditions for the cogeneration system were EUF=0.851 for HPR=2.2. HPR was varied in increments of 0.2 to simulate decreased heating requirements
- As the amount of heat needed onsite decreased, EUF adjusted to maintain a constant ratio of EUF/(1+HPR)=0.266. This represents heat that is available from the power generation, but not utilized

Figure 2 shows the results for an offsite power system efficiency of 0.263 compared to the cogeneration case, over a range of heating requirements, as represented by the HPR. The plots include CO_2 production rates for the two alternatives, and the emission ratio, the ratio of offsite to cogeneration CO_2 production (CO_2 production for the offsite system includes both power plant and onsite emissions from the local fuel based heating system). For these conditions, the cogeneration system produces less CO_2 over the full range, and is better by a factor of greater than 1.7 for the highest HPR. This is because the cogeneration system is sized to provide all the heat required at full electrical power production, so all of that heat is essentially free, while the offsite power system relies on the local heating system to provide the required heat, so its CO_2 production increases with heating load.

Figure 3 shows similar results for three offsite power system efficiencies. It is clear that as system efficiency increases, or the carbon content of the fuel decreases, the offsite power generation produces less CO_2 at lower values of HPR than the cogeneration system (i.e. CO_2 ratio is less than one). This is because the cogeneration system has to maintain full output to meet the electrical power load, thereby generating excess heat, while the offsite/onsite heating system can reduce output to match the heating load. These results are just examples, since actual values would depend on the systems and conditions being compared, particularly the carbon content and relative efficiencies of the combustion processes.

While not part of this analysis, the addition of a heat storage capacity could improve the cogeneration performance, provided that there was a recurring need for thermal energy that could be provided by excess heat recovered from storage. There are many other ways that energy and emissions performance could be improved based on specific circumstances and operating characteristics.

While the above analysis used single values for system efficiency, in a real world case, it would be better to use an average value that was representative of the mix of electrical power sources feeding the grid at any point in time. The EMS could monitor this information using Internet resources, and make decisions based on current conditions, determining when it made sense to use the onsite capacity, and when it made sense to use the offsite capacity. These decisions could be based on any combination of energy, environmental and cost factors.

Conclusions

Energy management systems have great potential for reducing GHG emissions as well as energy usage by monitoring and manipulating the operation of energy consuming systems. There are many approaches that are available for taking advantage of the carbon characteristics of the energy sources to minimize the GHG emissions associated with a particular energy requirement. Utilizing cogeneration, heat recovery, fuel switching and energy storage can all be effective measures provided that the sensing and control capabilities are incorporated into the EMS, and suitable algorithms are available for comparing alternatives under dynamic conditions. Information is increasingly becoming available via the Internet to enable the determination of the most effective operating strategies as both source and site characteristics change with time.

This paper demonstrated a simple tool for evaluating the GHG impact of different methods for providing facility electrical power and process heat. The results suggest that in many cases, cogeneration can result in a significant reduction in GHG emissions, although the relative performance is strongly dependent on the energy conversion efficiencies and the carbon content of the energy sources.

References

1. IPCC, 2007: Climate Change 2007: Mitigation. Contribution of Working Group III to the Fourth Assessment Report of the Intergovernmental Panel on Climate Change, Cambridge University Press, Cambridge, UK, and New York, USA.
2. Capehart, B.L., Turner, W.C., Kennedy, W.J., Guide to Energy Management, 5th Edition, The Fairmont Press, 2006.
3. Haines, R.W., Hittle, D.C., Control Systems for Heating, Ventilating and Air Conditioning, 6th Edition, Kluwer Academic Publishers, 2003.
4. Sorensen, B., Renewable Energy, Third Edition, Elsevier Academic Press, Burlington, MA, 2004.
5. Khartchenko, N. V., Advanced Energy Systems, Taylor and Francis, Washington, DC, 1998.

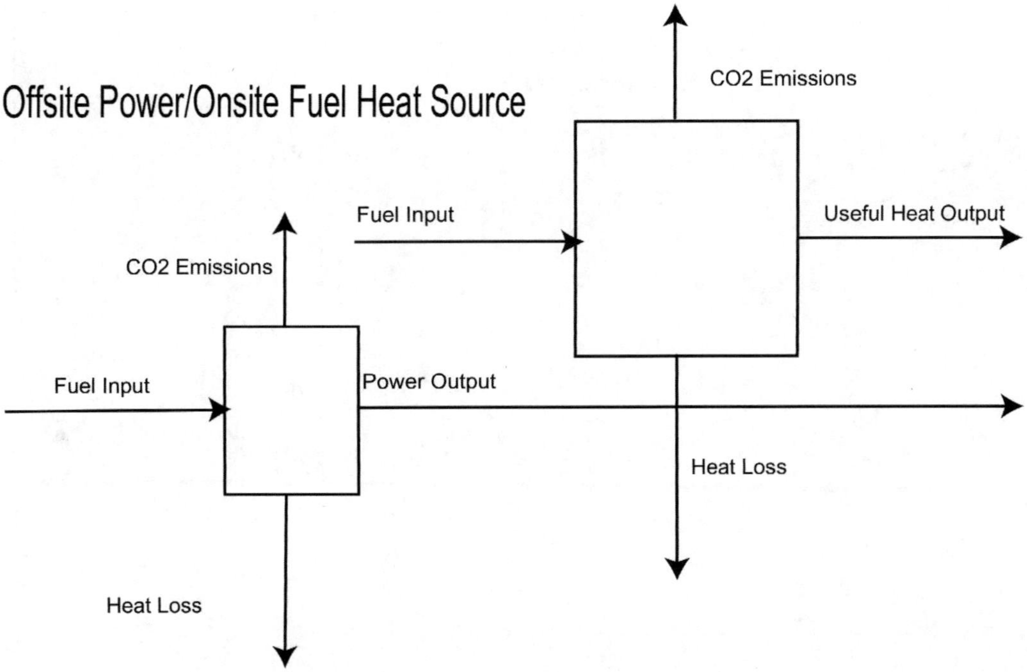

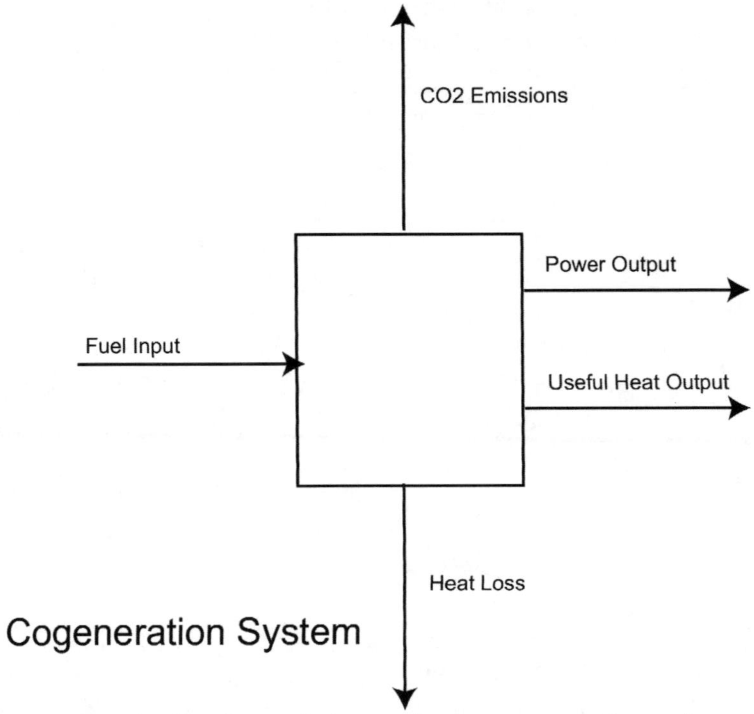

Figure 1. Schematic diagram of alternative system configurations for providing electrical power and heating energy

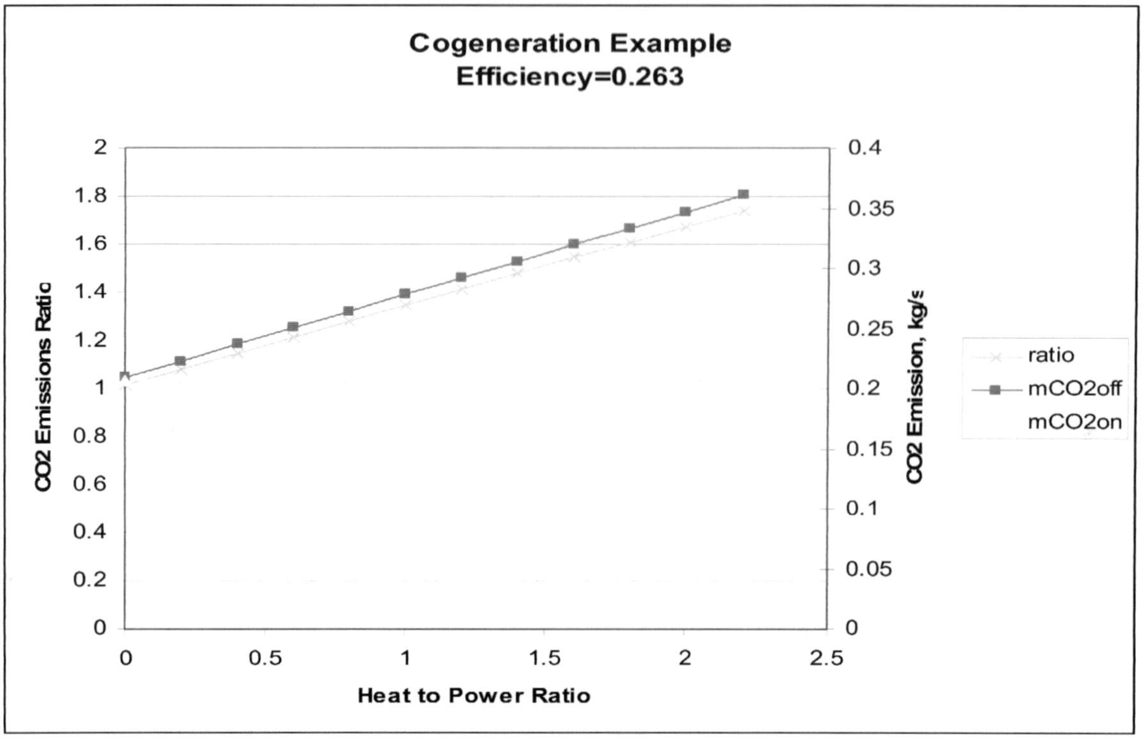

Figure 2. CO_2 emissions and emission ratio for offsite power generation with onsite heating versus onsite cogeneration

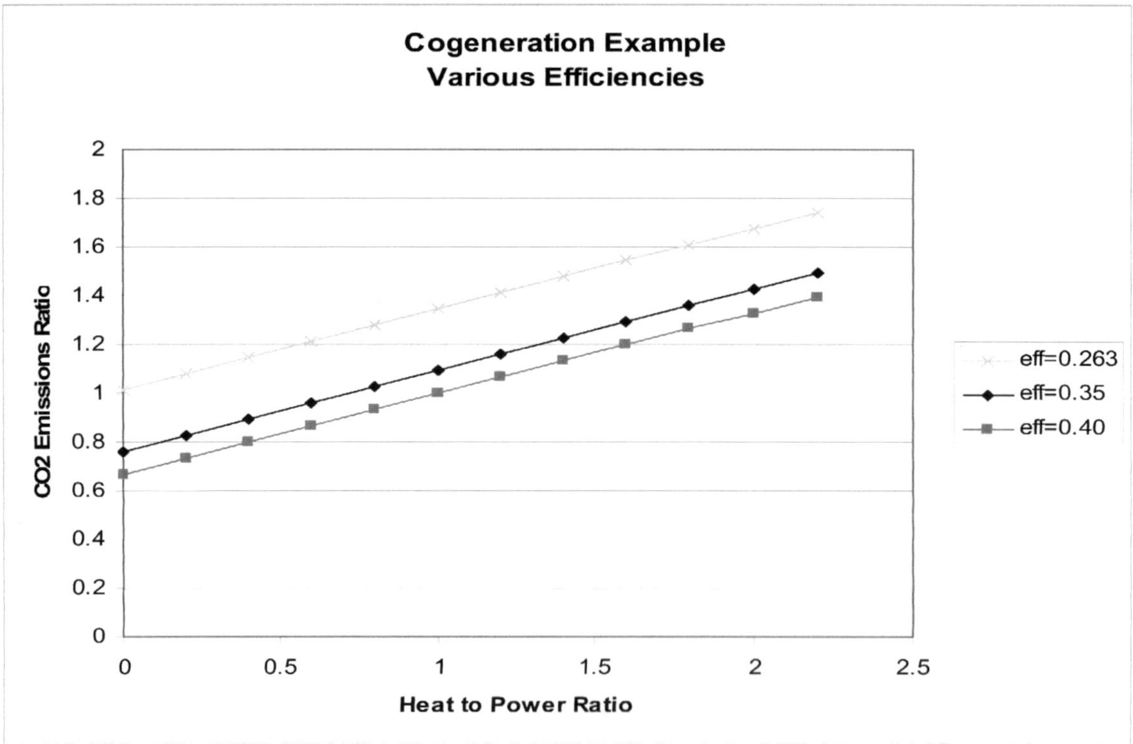

Figure 3. CO_2 emission ratios for offsite power generation with onsite heating versus onsite cogeneration for various offsite power generation efficiencies

Pump Load Management Via Repeated Simulation

Majid Al-Gwaiz[1], **Ahmed Aseeri**[1] and **Mahmoud Bahy Noureldin**[2],
(1)consulting services, Saudi Aramco, Dhahran, Saudi Arabia, (2)Consulting Services Department, Saudi Aramco, Dhahran, Saudi Arabia

Pumping systems are considered among the largest energy consumers in many production facilities. As such, operations optimization of these systems is an important ingredient of a successful energy efficiency optimization effort. Some industries nowadays do have offline pump load management advisory systems that are based on analyzing predefined operating scenarios. However, the variability of a process and its effect on a pumping network's dynamics can make it difficult to predict and analyze every operating condition and to come up with optimal network settings beforehand, which can result in missed energy optimization opportunities. Online optimization of pumping systems operations can capture some of these missed opportunities, and as a result, an estimated saving of up to 3% in the energy/electricity consumption of big pumping systems can be attained.

The pumping systems on-line optimization problem addressed in this paper is a typical supply-demand constrained optimization problem under uncertainty. This paper addresses a modeling and optimization method that can be applied to a variety of configurations, including series, parallel, and hybrid systems, and can also account for the settings of controllable equipment such as control valves and adjustable speed drives.

A number of optimization formulations have been considered for the problem at hand. The initial approach used a mixed integer nonlinear program that required some remodeling to eliminate the non-linearity from the integer variables, but the resulting model required advanced solvers and did not always converge to the global optimal. The second attempt used linearization techniques to transform the problem into a mixed integer linear program, but the size of the model grew rapidly. A repeated simulation approach was finally used and has shown to give reliable results even using simple solvers such as MS Excel's standard solver.

AUTHOR INDEX

AUTHOR INDEX